INTRODUCTION TO ISOTOPE HYDROLOGY

INTERNATIONAL CONTRIBUTIONS TO HYDROGEOLOGY

25

INTERNATIONAL ASSOCIATION OF HYDROGEOLOGISTS

Water
Resources
Programme

Introduction to Isotope Hydrology

Stable and Radioactive Isotopes of Hydrogen, Oxygen and Carbon

Willem G. Mook
Centre for Isotope Research, Groningen University, The Netherlands

Taylor & Francis
Taylor & Francis Group

LONDON / LEIDEN / NEW YORK / PHILADELPHIA / SINGAPORE

BALKEMA - Proceedings and Monographs
in Engineering, Water and Earth Science

© 2006 Taylor & Francis Group, London, UK

Typeset in Times New Roman by Newgen Imaging Systems (P) Ltd,
Chennai, India
Printed and bound in Great Britain by TJ International Ltd, Padstow,
Cornwall

Published by: Taylor & Francis/Balkema
                P.O. Box 447, 2300 AK Leiden, The Netherlands
                e-mail: Pub.NL@tandf.co.uk
                www.balkema.nl, www.tandf.co.uk, www.crcpress.com

British Library Cataloguing in Publication Data
A catalogue record for this book is available from the British Library

Library of Congress Cataloging in Publication Data
A catalog record for this book has been requested

ISBN10 0–415–38197–5   ISBN13 9–78–0–415–38197–0 (Hbk)

CONTENTS

LIST OF FIGURES

LIST OF TABLES

FOREWORD

As the scope of water challenges at local, regional and global levels continues to grow, it is ever more important to gain information for understanding the water cycle and hydrological systems. In recognition of this, the International Atomic Energy Agency (IAEA), through its Water Resources Programme, and UNESCO through its International Hydrological Programme (IHP), have long recognized the need for international co-operation on scientific research, education and training in the field of Hydrology. The IAEA has been working for several decades to promote and develop the use of isotopic techniques in hydrological research and, within the framework of the ongoing IHP-VI phase (2002–07), has established a close co-operation with UNESCO on the specific aspects of scientific and methodological developments related to nuclear techniques in Hydrology and water resources management.

A large part of the IAEA's support to its Member States, and to the water community as a whole, is through activities, publications, etc. to facilitate training and capacity building. To reflect this, the IAEA and the UNESCO International Hydrological Programme initiated and supported publication in 2000 of the book series *Environmental Isotopes in the Hydrological Cycle*.

The present volume represents a valuable contribution to the efforts of the UN Organizations towards a better knowledge of the hydrological cycle, and the critical gap for an introductory text needed for Isotope Hydrology courses and training activities is now being filled. This provides another important step to making the fundamental concepts of Isotope Hydrology more accessible to those students, teachers, water professionals, etc. who need them.

IAEA, Vienna, Austria   UNESCO, Paris, France
May 2005

PREFACE

In the year 2000 a series of textbooks on 'Environmental Isotopes in the Hydrological Cycle' were published, initiated by the International Atomic Energy Commission in Vienna and UNESCO in Paris, and written by a number of specialists in the field. The preface to this publication stated:

'The availability of freshwater is one of the great issues facing mankind today – in some ways the greatest, because problems associated with it affect the lives of many millions of people. It has consequently attracted widescale international attention of UN Agencies and related international/regional governmental and non-governmental organisations. The rapid growth of population coupled to steady increase in water requirements for agricultural and industrial development have imposed severe stress on the available freshwater resources in terms of both the quantity and quality, requiring consistent and careful assessment and management of water resources for their sustainable development.

More and better water cannot be acquired without the continuation and extension of hydrological research. In this respect the development and practical implementation of isotope methodologies in water resources assessment and management has been part of the International Atomic Energy Agency's (IAEA) programme in nuclear applications over the last four decades. Isotope studies applied to a wide spectrum of hydrological problems related to both surface and groundwater resources, as well as environmental studies in hydro-ecological systems, are presently an established scientific discipline often referred to as 'Isotope Hydrology'. The IAEA contributed to this development through direct support to research and training, and to the verification of isotope methodologies through field projects implemented in Member States.

The world-wide programme of the International Hydrological Decade (1965–74) and the subsequent long-term International Hydrological Programme (IHP) of UNESCO have been an essential part of the well recognised international frameworks for scientific research, education and training in the field of hydrology. The IAEA and UNESCO have established a close co-operation within the framework of both the earlier IHD and the ongoing IHP in the specific aspects of scientific and methodological developments related to water resources that are of mutual interest to the programmes of both organisations.'

The series of six up-to-date textbooks was published by UNESCO International Hydrological Programme as *Technical Document in Hydrology 39* and contained almost 1000 pages of basic treatment of the various aspects of the principles and applications of natural isotopes in water research. The main aim of the series was to provide a comprehensive review of basic theoretical concepts and principles of isotope hydrology methodologies and their practical applications with some illustrative examples. The volumes were designed to serve as 'Teaching Material' or textbooks for universities and teaching institutions and, furthermore, to be self-sufficient reference material for scientists and engineers involved in research and/or practical applications of isotope hydrology as an integral part of investigations related to water resources assessment, development and management.

The six volumes were prepared through efforts and contributions by a number of scientists involved in this specific field as cited in each volume, under the co-ordination of the authors designated for each volume, consecutively: W.G. Mook (Netherlands), J. Gat (Israel), K.Rozanski (Poland), K. Fröhlich (Austria), M. Geyh (Germany), K.P. Seiler (Germany) and Y. Yurtsever (IAEA, Vienna). Final editorial work on all volumes, aiming to achieve consistency in the contents and layout throughout the whole series, was allocated to the first author.

It is now felt desirable to have a concise volume that can serve as an introductory textbook for students, trainees, laboratory assistants and those hydrologists who become involved in projects including the application of isotope techniques. It is less extensive than the existing series but more complete than the first volume alone which was meant to cover the physical and chemical background of the occurrence of isotopes in nature. The material used is essentially extracted from the extensive UNESCO/IAEA series. It contains the majority of introductory chapters of Volume I, with the addition of some chapters from Volumes II and III, and is completed by material from lecture notes of the author at the Amsterdam Free University. All texts are prepared by the present author. In order to enlarge the readability, the long list of references in the text has been reduced to contain only those that are the most basic and of historic value. In case the reader is interested in the details, the original papers can be found in the original volumes, available as pdf files on internet site www.iaea.org/programmes/ripc/ih/volumes/volumes.htm.

The original studies on isotopes in water were concerned with seawater and precipitation. The first was primarily a survey on variations in $^{18}O/^{16}O$ concentration ratios, soon to be followed by a study of the $^{2}H/^{1}H$ ratios in natural waters (Friedman, 1953). Dansgaard (1964) observed $^{18}O/^{16}O$ variations in global precipitation in great detail, including a discussion on the meteorological patterns. His work was the start of the global 'isotopes-in-precipitation' network of the international organisations WMO and IAEA. In more recent years the observations were supported by theoretical and numerical modelling.

The first study on ^{14}C in groundwater, soon in combination with $^{13}C/^{12}C$, was initiated by the Heidelberg group in the late 1950s (Münnich, 1957; Vogel and Ehhalt, 1963). In later years this methodology became an important tool in studying groundwater movement.

The revolutionary development caused by the introduction of nuclear accelerators as mass spectrometers has greatly stimulated the hydrological application of isotopes with extremely low abundances in nature. ^{14}C research has also gained tremendously by this new technological approach.

The nature of isotopic applications is dictated by the specific character of isotopes, radioactive or non-radioactive, distinguishing three different types of application:

1. Stable and radioactive isotopes can be used as *tracers*, marking a water body or a certain quantity of water; an example is the phenomenon that rain water during a heavy storm is often depleted in the heavy isotope (stable ^{2}H, deuterium or stable ^{18}O) with respect to the most abundant isotope (^{1}H and ^{16}O, respectively). This offers the possibility to follow rain water in surface runoff, and even quantitatively analysing the runoff hydrograph.
2. During the transition of compounds such as water or carbon dioxide from one phase to another the concentration ratio of the isotopes of an element often changes,

and undergoes so-called *isotope fractionation*. Conversely, observing differences in especially the stable isotopic concentration ratios informs us about certain geochemical or hydrological processes that took place. For instance, as a result of a series of processes, the isotopic composition of carbon as well as oxygen for calcium carbonate is different for marine and freshwater origins. Furthermore, the isotopic composition of oxygen and hydrogen in rainwater varies with latitude, altitude, climate and the time of year.

3. *Radioactive decay* offers the possibility to determine an age, provided certain conditions are met. Noteworthy in this respect is the frequent application of dating groundwater – that is, determining the time elapsed since the infiltration of the water – by comparing the ^{14}C or ^{3}H (tritium) activities in a groundwater sample with that of the recharge water. Moreover, concentration differences of radioactive isotopes can also be used as a tracer.

In addition to the occurrence of natural isotopes in our environment, man is able to produce radioactive isotopes. These can also be applied as tracers in order to follow water movement or reservoir leakage. The methods of applying *artificial tracers* are parallel to the use of chemical tracers, often recognisable by their fluorescent character. These are not the subject of this volume despite the many successful applications.

W.G. Mook
Groningen, 1 April 2005

CHAPTER 1

Atomic systematics and nuclear structure

The principles and systematics of atomic and nuclear physics are summarised briefly in this chapter, in order to introduce the existence and characteristics of isotopes.

1.1 ATOMIC STRUCTURE AND THE PERIODIC TABLE OF THE ELEMENTS

Atoms consist of a nucleus surrounded by electrons. Compared to the diameter of an atom, which is in the order of 10^{-8} cm, the size of the nucleus is extremely small ($\sim 10^{-12}$ cm). The dense concentration of matter of the nucleus consists mainly of two kinds of particles, neutrons and protons, which have about the same mass. The neutron carries no electric charge, while the proton is positively charged. The number of protons (Z), the *atomic number*, is equal to the number of electrons surrounding the nucleus. Electrons have a mass that is about 1/1800 that of the proton mass and carry an equal but negative electrical charge, so that the atom as a whole is neutral. Atoms missing one or more electrons are referred to as positive ions, and atoms with a number of electrons exceeding the atomic number are called negative ions.

Protons and neutrons, the building stones of the nucleus, are called *nucleons*. The sum of the number of protons and neutrons (N) in a nucleus is the nuclear *mass number*:

$$A = Z + N \tag{1.1}$$

The notation describing a specific nucleus (= *nuclide*) of element X is:

$$^A_Z X_N$$

Because the chemical properties of an element (X) are primarily determined by the number of electrons in the atom, the atomic number Z characterises the element. Therefore, writing $^A X$ alone defines the nuclide. The cloud of electrons circulating around the nucleus is well structured and consists of shells, each containing a maximum number of electrons. The chemical properties of an atom are now mainly determined by the number of electrons in the outer, incompletely filled electron shell. Because of this systematic, all atoms can be arranged in a *Periodic Table of the Elements* (part shown by Fig. 1.1).

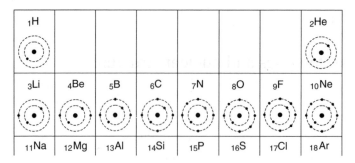

Figure 1.1.   Part of the Periodic Table of the Elements, containing the light elements. Also shown are the electronic configurations of the respective atoms.

1.2   STRUCTURE OF THE ATOMIC NUCLEUS

The atomic nuclei are held together by extremely strong forces between the nucleons (protons and neutrons) with a very small range. As repulsive electrical (Coulomb) forces exist between the protons, the presence of neutrons is required to stabilise the nucleus. In the most abundant nuclides of the *light elements*, the numbers of protons and neutrons are equal. Nuclei such as

$$\quad {}^{2}_{1}H_{1} \qquad {}^{4}_{2}He_{2} \qquad {}^{12}_{6}C_{6} \qquad {}^{14}_{7}N_{7} \qquad {}^{16}_{8}O_{8}$$

are stable, as is the single proton (^{1}H = hydrogen). For the heavy elements the number of neutrons far exceeds the number of protons: ^{238}U contains only 92 protons, whereas the largest stable nuclide, the lead isotope ^{208}Pb, has an atomic number of 82.

Instabilities are caused by an excess of protons or neutrons. Examples of such unstable or *radioactive* nuclei are

$$\quad {}^{3}_{1}H_{2} \quad \text{and} \quad {}^{14}_{6}C_{8}$$

For the light elements, a slight excess of neutrons does not necessarily result in unstable nuclei:

$$\quad {}^{13}_{6}C_{7} \qquad {}^{15}_{7}N_{8} \qquad {}^{17}_{8}O_{9} \qquad {}^{18}_{8}O_{10}$$

are stable. For these 'asymmetric' nuclei ($Z \neq N$) however, the probability of formation during the 'creation' of the elements – nucleosynthesis – was smaller, resulting in smaller natural concentrations for these nuclides.

The naturally occurring stable and radioactive isotopes of the light elements are shown in a part of the *Chart of Nuclides* (Fig. 1.2). In the higher A range, for instance with the radioactive decay series (Section 4.2.7), N and Z are no longer about equal (^{238}U with $Z = 92$ and $N = 146$).

1.3   STABLE AND RADIOACTIVE ISOTOPES

The atomic nuclei of an element containing different numbers of neutrons are called *isotopes* (ισο τοπος = at the same place in the Periodic Table of the Elements). Many elements

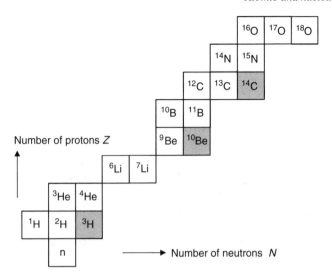

Figure 1.2.   Part of the Chart of Nuclides, containing the light elements. The isotopes of an element (equal Z) are found in a horizontal row, isobars (equal A) along diagonal lines, isotones (equal N) in vertical columns. The natural radioactive isotopes of H, Be and C are marked grey.

have two or more stable, naturally occurring isotopes. In general, nuclei with even numbers of protons or/and neutrons are more stable. Nuclei of which the number of protons or/and neutrons corresponds to some specific even number, belonging to the series of so-called *magic numbers* 2, 8, 20, 28, 50, 82 and 126, have a relatively high stability and consequently large natural occurrence.

These magic numbers can be explained by a nuclear shell model with *closed nucleon shells* within the nucleus, similar to the closed electron shells in atoms, the basis of the periodicity in the Periodic Table of the Elements. An example of the occurrence of the magic numbers is the large number of stable isotopes of lead: the largest stable nuclide, ^{208}Pb (with $Z = 82$ and $N = 126$) is double magic. The uneven–uneven nuclei are especially unstable and have a small chance of natural occurrence. Most uneven-Z elements have only one or at the most two stable isotopes.

The earlier statement that the chemical properties of an element only depend on the atomic number, implying that the chemical properties of isotopes are equal, needs revision if we look in detail. The fact is that in nature variable relative concentrations of isotopes are observed. There are two causes for this phenomenon:

1. The chemical and some physical properties of the isotopes of one element are not exactly equal, resulting in slightly different chemical and physical properties – and consequently different concentrations – of isotopic molecules, that is, molecules that contain different isotopes of that element.
2. If the isotopes concerned are radioactive, the process of radioactive decay causes the concentration of the isotopic molecules concerned to diminish in time; this may result in concentration differences that are much larger than those caused by the isotope processes mentioned under 1.

These phenomena will be discussed separately in later chapters.

1.4   MASS AND ENERGY

It is inconvenient to use the real mass of atoms and molecules. Instead we define the *atomic mass* as the mass expressed in atomic mass units (amu). Originally this was equivalent to the mass of a proton; later for practical reasons, the amu has rather become defined as 1/12 times the mass of a ^{12}C atom:

$$1 \text{ amu} = 1.6605655 \times 10^{-27} \text{ kg} \tag{1.2}$$

In chemistry it is now convenient to use the *mole* quantity, defined as the number of grams of the element equal to the atomic weight. The number of atoms in one mole of the element, or the number of molecules in one mole of a chemical compound, is then equal for any substance and is given by Avogadro's number $= 6.02252 \times 10^{23}$.

If we now add up the atomic masses of the building stones of a certain nucleus X, for instance of ^{12}C consisting of 6 protons and 6 neutrons, it appears that the atomic mass of X is smaller than the sum of the 12 constituent particles:

6 protons $= 6 \times 1.007825$ amu	$= 6.04695$ amu
6 neutrons $= 6 \times 1.008665$ amu	$= 6.05199$ amu
Total mass	$= 12.09894$ amu
Compared to: A for ^{12}C	$= 12.00000$ amu (by definition)
Difference	$= 0.09894$ amu

This so-called *mass defect* has been converted into the *binding energy*, the potential energy 'stored' in the nucleus to keep the particles together. This equivalence of mass and energy is defined by Einstein's special theory of relativity as

$$E_B = Mc^2 \tag{1.3}$$

where E_B is the binding energy, M is the mass and c the velocity of light (2.997925×10^8 m/s).

According to this definition the *mass-energy equivalence* is expressed as

$$1 \text{ amu} \equiv 931.5 \text{ MeV (million electronvolt)} \tag{1.4}$$

where $1 \text{ eV} = 1.602189 \times 10^{-19}$ J (electrical charge unit = electron charge = 1.602189×10^{-19}C).

All energies in nuclear physics, such as the particle energies after nuclear decay, are presented in MeV or keV (kilo or 10^3 electronvolt). In our case of ^{12}C the binding energy turns out to be 0.09894 amu $\times 931.5$ MeV/amu $= 92.16$ MeV or 7.68 MeV per nucleon.

CHAPTER 2

Isotope fractionation

In classical chemistry, isotopes of an element are regarded as having equal chemical properties. In reality, variations in isotopic abundances occur which far exceed measuring precision. This phenomenon is the subject of the present chapter.

2.1 ISOTOPE RATIOS AND CONCENTRATIONS

Before we can give a more quantitative description of *isotope effects*, we must define isotope abundances more carefully. *Isotope (abundance) ratios* are defined by the expression

$$R = \frac{\text{abundance of rare isotope}}{\text{abundance of abundant isotope}} \tag{2.1}$$

> The ratio carries a superscript before the ratio symbol R, which refers to the isotope under consideration. For instance:

$$^{13}R(CO_2) = \frac{[^{13}CO_2]}{[^{12}CO_2]} \quad ^{18}R(CO_2) = \frac{[C^{18}O^{16}O]}{[C^{16}O_2]}$$

$$^{2}R(H_2O) = \frac{[^{2}H^{1}HO]}{[^{1}H_2O]} \quad ^{18}R(H_2O) = \frac{[H_2{}^{18}O]}{[H_2{}^{16}O)} \tag{2.2}$$

We should clearly distinguish between an *isotope ratio* and an *isotope concentration*. For CO_2, the latter is defined by:

$$\frac{[^{13}CO_2]}{[CO_2]} = \frac{[^{13}CO_2]}{[^{13}CO_2] + [^{12}CO_2]} = \frac{^{13}R}{1 + {}^{13}R} \tag{2.3}$$

If the rare isotope concentration is very large, as in the case of labelled compounds, the rare isotope concentration is often given in *atom %*. This is then related to the isotope ratio by:

$$R = (\text{atom \%}/100)/[1 - (\text{atom \%}/100)] \tag{2.4}$$

2.2 ISOTOPE FRACTIONATION

According to classical chemistry, the chemical characteristics of isotopes, or rather of molecules that contain different isotopes of the same element (such as $^{13}CO_2$ and $^{12}CO_2$) are equal. To a large extent this is true. However, if a measurement is sufficiently accurate – and this is the case with modern analytical equipment (Appendix II) – we observe tiny differences in chemical as well as physical behaviour of so-called *isotopic molecules* or *isotopic compounds*. The phenomenon that these isotopic differences exist is called *isotope fractionation*. (Some authors refer to this phenomenon as *isotope discrimination* or *isotope separation*; however, we see no reason to deviate from the original expression.) This can occur as a change in isotopic composition by the transition of a compound from one state to another (liquid water to water vapour) or into another compound (carbon dioxide into plant organic carbon), or it can manifest itself as a difference in isotopic composition between two compounds in chemical equilibrium (dissolved bicarbonate and carbon dioxide) or in physical equilibrium (liquid water and water vapour). Examples of all these phenomena will be discussed throughout this volume.

The differences in physical and chemical properties of isotopic compounds (i.e. chemical compounds consisting of molecules containing different isotopes of the same element) are brought about by mass differences of the atomic nuclei. The consequences of these mass differences are two-fold:

1. The heavier isotopic molecules have a lower mobility. The kinetic energy of a molecule is solely determined by temperature: $kT = \frac{1}{2}mv^2$ (k = Boltzmann constant, T = absolute temperature, m = molecular mass, v = average molecular velocity). Therefore, molecules have the same $\frac{1}{2}mv^2$, regardless of their isotope content. This means that the molecules with larger m necessarily have a smaller v. Some practical consequences are:

   (a) heavier molecules have a lower diffusion velocity;
   (b) the collision frequency with other molecules – the primary condition for chemical reaction – is smaller for heavier molecules; this is one of the reasons why, as a rule, lighter molecules react faster.

2. The heavier molecules generally have higher *binding energies* – that is, the kinetic energy of any two atoms within a molecule, or any two molecules of a certain compound needed to overcome the attractive forces between them. A simple example is the heat of evaporation: $^1H_2{}^{18}O$ and $^1H^2H^{16}O$ have lower vapour pressures than $^1H_2{}^{16}O$; therefore, they evaporate less easily. In other words, the vapour pressure for the isotopically heavy species is lower (*normal isotope effect*). On the contrary, *inverse isotope effects* can occur, where the isotopically heavy particles have the higher vapour pressure. Examples are the higher vapour pressure of $^{13}CO_2$ in the liquid phase, as well as the lower solubility of $^{13}CO_2$ in water than $^{12}CO_2$.

2.3   KINETIC, EQUILIBRIUM AND NON-EQUILIBRIUM ISOTOPE FRACTIONATION

2.3.1   Definitions

The process of isotope fractionation is mathematically described by comparing the isotope ratios of the two compounds in chemical equilibrium (A ⇔ B) or of the compounds before and after a physical or chemical transition process (A → B). The *isotope fractionation factor* is then defined as the ratio of the two isotope ratios:

$$\alpha_A(B) = \alpha_{B/A} = \frac{R(B)}{R(A)} = \frac{R_B}{R_A} \tag{2.5}$$

which expresses the isotope ratio in the phase or compound B relative to that in A.

If we are dealing with changes in isotopic composition, for instance when C is oxidised to CO_2, the carbon isotope fractionation refers the 'new' $^{13}R(CO_2)$ value to the 'old' $^{13}R(C)$; in other words $^{13}\alpha = {}^{13}R(CO_2)/{}^{13}R(C)$.

In general, isotope effects are small: $\alpha \approx 1$. Therefore, the deviation of α from 1 is widely used rather than the *fractionation factor*. This quantity, to which we refer as the *fractionation*, is defined by:

$$\varepsilon_{B/A} = \alpha_{B/A} - 1 = \frac{R_B}{R_A} - 1 \quad (\times 10^3\text{‰}) \tag{2.6}$$

ε represents the *enrichment* ($\varepsilon > 0$) or the *depletion* ($\varepsilon < 0$) of the rare isotope in B with respect to A. The symbols $\alpha_{B/A}$ and $\varepsilon_{B/A}$ are equivalent to $\alpha_A(B)$ and $\varepsilon_A(B)$. In the one-way process (A → B) ε is the change in isotopic composition; in other words, the new isotopic composition compared to the old.

Because ε is a small number, it is generally given in ‰ (per mill, equivalent to 10^{-3}). Note that *we do not define ε in ‰*, as many authors claim they do (in fact they do not, as they always add the ‰ symbol). An ε value of, for example, 5‰ is equal to 0.005. The consequence is that *in mathematical equations it is incorrect to use $\varepsilon/10^3$ instead of merely ε*. The student is reminded that ε is a small number; in equations one may numerically write, for instance, '−25‰' instead of '−25/10³'.

Again, the fractionation of B with respect to A is denoted by $\varepsilon_{B/A}$ or $\varepsilon_A(B)$. From the definition of ε we simply derive:

$$\varepsilon_{B/A} = \frac{-\varepsilon_{A/B}}{1 + \varepsilon_{A/B}} \approx -\varepsilon_{A/B} \tag{2.7}$$

the last approximation because, in natural processes, the ε values are small.

It is important to distinguish between three kinds of isotope fractionation (Fig. 2.1):

1. kinetic fractionation
2. equilibrium fractionation and
3. non-equilibrium fractionation.

Kinetic fractionation results from *irreversible* that is, *one-way* physical or chemical processes; in other words, there is a one-way mass transport. Examples are the combustion

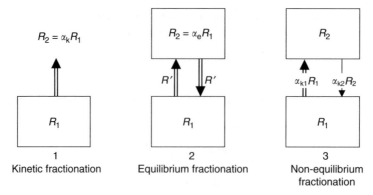

Figure 2.1.   Three modes of isotope fractionation: (1) *kinetic fractionation* (= α_k), with one-way mass transport; (2) *equilibrium fractionation* (= α_e), without a net mass transport; the stationary state of the isotopic compositions of both compounds requires the transfer of equal masses with equal isotopic compositions (= R'); (3) *non-equilibrium fractionation*, with a net transfer of masses between the two compounds, which are both kinetically fractionated.

of organic matter, or the evaporation of water with immediate withdrawal of the vapour from further contact with the water, or the absorption and diffusion of gases, or the bacterial decay of plants or rapid calcite precipitation. These fractionation effects are primarily determined by the binding energies of the original compounds (Section 2.2): during physical processes isotopically lighter molecules have higher velocities and smaller binding energies; in chemical processes light molecules react more rapidly than the heavy. Rarely, however, is the opposite true: the *inverse kinetic isotope effect* occurs most commonly in reactions involving hydrogen atoms.

The second type of fractionation is *equilibrium* (or *thermodynamic*) *fractionation*. This is essentially the isotope effect involved in a *reversible* (thermodynamic) equilibrium reaction, where the total as well as isotope mass transport is equal in both directions. Examples are the equilibrium between water liquid and water vapour in a closed volume, or the stationary exchange between dissolved CO_2 and HCO_3^- in water.

2.3.2   Isotope exchange

As a formal example of equilibrium fractionation we choose the isotope exchange reaction:

$$^*A + B \Leftrightarrow A + {}^*B \tag{2.8}$$

where the asterisk indicates the presence of the rare isotope. The fractionation factor for this equilibrium between phases or compounds A and B is the equilibrium constant for the exchange reaction of Eq. 2.8:

$$K = \frac{[A]}{[^*A]}\frac{[^*B]}{[B]} = \frac{[^*B]/[B]}{[^*A]/[A]} = \frac{R_B}{R_A} = \alpha_{B/A} \tag{2.9}$$

If sufficient information about the binding energies of atoms and molecules is available, the fractionation effect, the kinetic effect as well as the equilibrium effect can be calculated. In practice, however, these data are often not known in sufficient detail. With kinetic isotope effects we are confronted with an additional difficulty which arises from the fact that natural processes are often not purely kinetic or irreversible. Moreover, kinetic fractionation is difficult to measure in the laboratory because: (i) complete irreversibility cannot be guaranteed (part of the water vapour will return to the liquid), nor can the degree of irreversibility be quantified; (ii) the vanishing phase or compound will have a non-homogeneous and often immeasurable isotopic composition, because the isotope effect occurs at the surface of the compound. For example, the surface layer of an evaporating water mass may become enriched in ^{18}O and ^{2}H if mixing within the water mass is not sufficiently rapid to keep its content homogeneous throughout. Also, the water content of plant leaf is more enriched in ^{2}H and ^{18}O than the total plant water.

Equilibrium fractionation, on the other hand, can be determined by laboratory experiments, and in several cases reasonable agreement has been shown between experimental data and thermodynamic calculations.

In nature, isotope fractionation processes are often not purely kinetic (i.e. one-way processes), and neither purely equilibrium processes (no net mass transport). These are referred to as *non-equilibrium fractionations*. An example is the evaporation of ocean or fresh surface water bodies; the evaporation is not a one-way process (since water vapour condenses), and it is not an equilibrium process (as there is a net evaporation).

2.3.3   Relation between equilibrium and kinetic fractionation

The general condition for the establishment of isotopic equilibrium between two compounds is the existence of an isotope exchange mechanism. This can be a reversible chemical equilibrium such as:

$$H_2{}^{16}O + C^{18}O^{16}O \Leftrightarrow H_2C^{18}O^{16}O_2 \Leftrightarrow H_2{}^{18}O + C^{16}O_2 \qquad (2.10)$$

or a reversible physical process such as evaporation/condensation:

$$H_2{}^{16}O(\text{vapour}) + H_2{}^{18}O(\text{liquid}) \Leftrightarrow H_2{}^{18}O(\text{vapour}) + H_2{}^{16}O(\text{liquid}) \qquad (2.11)$$

The reaction rates of exchange processes and, consequently, the periods of time required to reach isotopic equilibrium, vary greatly. For example, the exchange of $H_2O \Leftrightarrow CO_2$ proceeds on a scale of minutes to hours at room temperature, while that of $H_2O \Leftrightarrow SO_4^{2-}$ requires millennia.

The fractionation resulting from kinetic isotope effects generally exceeds that from equilibrium processes. Moreover, in a kinetic process the compound formed may be depleted in the rare isotope, while it is enriched in the equivalent equilibrium process. This can be understood by comparing the fractionation factor in a reversible equilibrium with the kinetic fractionation factors involved in the two opposed single reactions. As an example, we take the carbonic acid equilibrium of Eq. 2.10:

$$CO_2 + H_2O \Leftrightarrow H^+ + HCO_3^-$$

For the single reactions

$$^{12}CO_2 + H_2O \rightarrow H^+ + H^{12}CO_3^-$$

and

$$^{13}CO_2 + H_2O \rightarrow H^+ + H^{13}CO_3^-$$

the reaction rates are, respectively:

$$^{12}r = {}^{12}k[^{12}CO_2] \quad \text{and} \quad {}^{13}r = {}^{13}k[^{13}CO_2] \tag{2.12}$$

where ^{12}k and ^{13}k are the reaction rate constants. The isotope ratio of the bicarbonate formed (ΔHCO_3^-) is:

$$^{13}R(\Delta HCO_3) = \frac{^{13}r}{^{12}r} = \frac{^{13}k[^{13}CO_2]}{^{12}k[^{12}CO_2]} = \alpha_k \, {}^{13}R(CO_2) \tag{2.13}$$

where α_k is the kinetic fractionation factor for this reaction. Conversely, for:

$$H^+ + H^{12}CO_3^- \rightarrow {}^{12}CO_2 + H_2O$$

and

$$H^+ + H^{13}CO_3^- \rightarrow {}^{13}CO_2 + H_2O$$

the reaction rates are

$$^{12}r' = {}^{12}k'[H^{12}CO_3^-] \quad \text{and} \quad {}^{13}r' = {}^{13}k'[H^{13}CO_3^-] \tag{2.14}$$

The carbon dioxide formed (ΔCO_2) has an isotope ratio

$$^{13}R(\Delta CO_2) = \frac{^{13}r'}{^{12}r'} = \frac{^{13}k'[H^{13}CO_3^-]}{^{12}k'[H^{12}CO_3^-]} = \alpha_k' \, {}^{13}R(HCO_3^-) \tag{2.15}$$

At isotopic equilibrium the isotope effects balance; in other words:

$$^{13}R(\Delta HCO_3^-) = {}^{13}R(\Delta CO_2) \tag{2.16}$$

so that combining Eqs 2.13 and 2.15 results in:

$$\frac{\alpha_k'}{\alpha_k} = \frac{^{13}R(CO_2)}{^{13}R(HCO_3^-)} = \frac{[^{13}CO_2][H^{12}CO_3^-]}{[^{12}CO_2][H^{13}CO_3^-]} = \alpha_e \tag{2.17}$$

which shows that the isotopic equilibrium fractionation α_e is equivalent to the equilibrium constant of the isotope exchange reaction:

$$H^{13}CO_3^- + {}^{12}CO_2 \Leftrightarrow H^{12}CO3 + {}^{13}CO_2 \tag{2.18}$$

Most α_k and α_k' values are less than 1 by more than one per cent. Differences between their ratios are less – in the order of several per mill. Later we will see that rapid evaporation of water might cause the water vapour to be about twice as depleted in ^{18}O as the vapour in equilibrium with the water. The reason for this is that $H_2{}^{16}O$ is favoured also in the condensation process.

From Eq. 2.17 it is obvious that, while α_k for a certain phase transition is smaller than 1, α_e might be larger than 1. An example is to be found in the system dissolved/gaseous CO_2: the ^{13}C content of CO_2 rapidly withdrawn from an aqueous CO_2 solution is smaller than that of the dissolved CO_2 (i.e. ^{12}C moves faster: $\alpha_k < 1$), while under equilibrium conditions the gaseous CO_2 contains relatively more ^{13}C ($\alpha_e > 1$; the inverse isotope effect, Section 2.2).

2.4 CONCLUSIONS FROM THEORY

In this section we present a summary of the characteristics of isotope effects, in particular the mass-dependent isotope fractionation, based on some principles from thermodynamics and statistical mechanics. For a full discussion reference should be made to the extensive UNESCO/IAEA volumes, specifically Volume I.

A general approximate expression for the fractionation factor as a function of the absolute temperature is:

$$\alpha = Ae^{B/T} \tag{2.19}$$

where the coefficients A and B do not depend on temperature, but contain all temperature-independent quantities (such as the atomic and molecular masses of the isotopes and molecules involved). The natural logarithm of the fractionation factor is approximated by the power series:

$$\ln \alpha = C_1 + \frac{C_2}{T} + \frac{C_3}{T^2} \tag{2.20}$$

with the often used approximation for the fractionation:

$$\varepsilon \approx \ln(1 + \varepsilon) = C_1 + C_2/T \tag{2.21}$$

At high temperatures the differences between binding energies – that is, the distances between the energy levels – of isotopic molecules become smaller, resulting in smaller and ultimately disappearing isotope effects (Fig. 2.1). In an isotope equilibrium between two chemical compounds the heavy isotope is generally concentrated in the compound which has the largest molecular weight.

In summary, these theoretical discussions lead to some general conclusions:

1. In a kinetic (one-way or irreversible) process the phase or compound formed is depleted in the heavy isotope with respect to the original phase or compound ($\alpha_k < 1$); theoretical predictions about the degree of fractionation can only be qualitative (rapidly evaporating water is depleted in ^{18}O relative to the water body itself).
2. In an isotopic equilibrium (reversible) process it cannot with certainty be predicted whether the one phase or compound is enriched or depleted in the heavy isotope. However, the dense phase (liquid rather than vapour) or the compound having the largest molecular mass ($CaCO_3$ versus CO_2), usually contains the highest abundance of the heavy isotope.

3. Under equilibrium conditions, and provided sufficient spectroscopic data on the binding energies are available, α_e may be calculated; measurements of α_e for various isotope equilibria are reported in Chapter 5.
4. As a rule fractionation decreases with increasing temperature. This is shown in Chapter 5 for the exchange equilibria of $CO_2 \Leftrightarrow H_2O$ and $H_2O_{liquid} \Leftrightarrow H_2O_{vapour}$. At very high temperatures the isotopic differences between the compounds disappear.

2.5   FRACTIONATION BY DIFFUSION

Isotope fractionation may occur because of the different mobilities of isotopic molecules. An example in nature is the diffusion of CO_2 or H_2O through air.

According to Fick's law the net flux of gas, F, through a unit surface area is:

$$F = -D\frac{dC}{dx} \tag{2.22}$$

where dC/dx is the concentration gradient in the direction of diffusion and D is the diffusion constant. The latter is proportional to the temperature and to $1/\sqrt{m}$, where m is the molecular mass. This proportionality results from the fact that all molecules in a gas (mixture) have equal temperature and thus equal average $\frac{1}{2}mv^2$. The average velocity of the molecules and, thus, their mobilities are inversely proportional to $\sqrt{m}$.

If the diffusion process of interest involves the movement of gas A through gas B, however, m has to be replaced by the reduced mass:

$$\mu = \frac{m_A m_B}{m_A + m_B} \tag{2.23}$$

(see textbooks on the kinetic theory of gases).

The given equations hold for the abundant as well as for the rare isotope. The resulting fractionation is then given by the ratio of the diffusion coefficient for the two isotopic species. Furthermore, the molecular masses can be replaced by the molar weights M in the numerator and denominator:

$$\alpha = \frac{{}^*D}{D} = \sqrt{\frac{{}^*M_A + M_B}{{}^*M_A \times M_B} \frac{M_A \times M_B}{M_A + M_B}} \tag{2.24}$$

In the example of water vapour diffusing through air, the resulting fractionation factor for oxygen is (M_B of air is taken to be 29, $M_A = 18$, ${}^*M_A = 20$):

$$^{18}\alpha = \left[\frac{20 + 29}{20 \times 29}\frac{18 \times 29}{18 + 29}\right]^{1/2} = 0.969 \tag{2.25}$$

By diffusion through air, water vapour thus will become depleted in ^{18}O (in agreement with the rules given at the end of Section 2.4) by 31 ‰: $^{18}\varepsilon = -31‰$.

The ^{13}C fractionation factor for diffusion of CO_2 through air is:

$$^{13}\alpha = \left[\frac{45 + 29}{45 \times 29}\frac{44 \times 29}{44 + 29}\right]^{1/2} = 0.9956 \tag{2.26}$$

in other words $^{13}\varepsilon = -4.4‰$, a depletion of 4.4 ‰.

2.6   RELATION BETWEEN ATOMIC AND MOLECULAR ISOTOPE RATIOS

We now wish to take a closer look at the meaning of the isotope ratio in the context of the abundance of rare isotopes of an element in poly-atomic molecules, containing at least two atoms of this element. The isotope ratio is the chance that the abundant isotope has been replaced by the rare isotope. To be more specific, we define the *atomic isotope ratio* (R_{atom}) as the abundance of all rare isotopic atoms in the compound divided by the abundance of all abundant isotopic atoms. For the sake of clarity, we select a simple example, the isotopes of hydrogen in water:

$$^2R_{atom} = \frac{[^2H^{16}O^1H] + [^2H^{17}O^1H] + [^2H^{18}O^1H] + 2[^2H^{16}O^2H] + \cdots}{2[^1H^{16}O^1H] + 2[^1H^{17}O^1H] + 2[^1H^{18}O^1H] + [^1H^{16}O^2H] + \cdots}$$

(2.27)

The 'second-order' abundances, such as for example $^2H_2{}^{18}O$ in the numerator and $^1H^2H^{18}O$ in the denominator, have been neglected. In this context the *molecular isotope ratio* (R_{mol}) is defined as the abundance of the rare isotope of a specific element in a specific type of molecule divided by the abundance of the abundant isotope. In our example of deuterium in water:

$$^2R_{mol} = \frac{[^2H^{16}O^1H]}{2[^1H^{16}O^1H]}$$

(2.28)

The realistic value of this definition refers to the measurement of isotope ratios by *laser (light) spectrometry* (see Appendix II), where the light absorption by two isotopic molecules, only differing in one isotope, is compared, contrary to *mass spectrometry*.

Eq. 2.28 can be rewritten as:

$$^2R_{atom} = \frac{[^2H^{16}O^1H]}{2[^1H^{16}O^1H]} \times \{(1 + [^2H^{17}O^1H]/[^2H^{16}O^1H]$$
$$+ [^2H^{18}O^1H]/[^2H^{16}O^1H] + 2[^2H^{16}O^2H]/[^2H^{16}O^1H] + \cdots)/$$
$$(1 + [^1H^{17}O^1H]/[^1H^{16}O^1H] + [^1H^{18}O^1H]/[^1H^{16}O^1H]$$
$$+ [^1H^{16}O^2H]/2[^1H^{16}O^1H] + \cdots)\}$$

(2.29)

The first factor on the right-hand side in Eq. 2.29 is the molecular isotope ratio. To a very good approximation, the molecular isotope ratios in the second factor may be replaced by the atomic isotope ratios:

$$^2R_{atom} \approx {}^2R_{mol} \frac{1 + {}^{17}R + {}^{18}R + {}^2R + \cdots}{1 + {}^{17}R + {}^{18}R + {}^2R + \cdots} = {}^2R_{mol}$$

(2.30)

This result means that to a first-order approximation the atomic isotope ratio is equal to a molecular isotope ratio. Although shown for only one specific molecule, the reasoning is equally valid for the other molecules in Eq. 2.29. Also, for more complicated molecules the demonstration of the proof is analogous.

The approximations made are indicated as 'first-order'. However, if we use δ values the approximation is even better, because almost equal approximations occur in the R values of the numerator as well as the denominator of Eq. 2.30 (cf. Section 3.2.1).

2.7　RELATION BETWEEN FRACTIONATIONS FOR THREE ISOTOPIC MOLECULES

For some isotope studies, it is of interest to know the relation between the fractionations for more than two isotopic molecules. For example, water contains three different isotopic molecules as far as oxygen is concerned: $H_2{}^{16}O$, $H_2{}^{17}O$ and $H_2{}^{18}O$; carbon dioxide consists of $^{12}CO_2$, $^{13}CO_2$ and $^{14}CO_2$. The question now is, whether there is a theoretical relation between the ^{17}O and ^{18}O fractionation with respect to ^{16}O (for instance, during evaporation) and between the ^{13}C and ^{14}C fractionation with respect to ^{12}C (e.g. during the uptake of CO_2 by plants).

Generally, the fractionation for the heaviest isotope is twice as large as that for the less heavy isotope:

$$^{14}\alpha = (^{13}\alpha)^2 \quad \text{and} \quad ^{18}\alpha = (^{17}\alpha)^2 \tag{2.31}$$

with

$$^{14}\alpha = 1 + {}^{14}\varepsilon = (1 + {}^{13}\varepsilon)^2 = 1 + 2{}^{13}\varepsilon + ({}^{13}\varepsilon)^2 \approx 1 + 2{}^{13}\varepsilon$$

so that for the CO_2 uptake by plants as well as the isotope exchange between CO_2 and water, respectively:

$$^{14}\varepsilon \approx 2{}^{13}\varepsilon \quad \text{and} \quad ^{18}\varepsilon \approx 2{}^{17}\varepsilon \tag{2.32}$$

or in general:

$$\frac{^{m+2}R_A}{^{m+2}R_B} = \left(\frac{^{m+1}R_A}{^{m+1}R_B}\right)^2 \quad \text{or} \quad \frac{\ln {}^{m+2}\alpha_{A/B}}{\ln {}^{m+1}\alpha_{A/B}} = 2 \tag{2.33}$$

where $m = 12$ for carbon or 16 for oxygen.

In other words:

$$\ln {}^{m+2}\alpha / \ln {}^{m+1}\alpha = 2 \quad \text{and} \quad ^{m+2}\alpha = (^{m+1}\alpha)^2 \tag{2.34}$$

or:

$$^{m+2}\varepsilon = 2^{m+1}\varepsilon \tag{2.35}$$

where $m = 12$ for carbon isotopes and $m = 16$ for oxygen.

2.8　USE OF δ VALUES AND ISOTOPE STANDARDS

Isotope ratios such as

$$^2R = \frac{^2H}{^1H} \qquad ^{13}R = \frac{^{13}C}{^{12}C} \qquad ^{18}R = \frac{^{18}O}{^{16}O} \tag{2.36}$$

are generally not reported as *absolute ratios*. The main reasons are:

1. The type of mass spectrometers, suitable for measuring isotope abundances with high sensitivity in order to detect very small natural variations, are basically not suitable for obtaining reliable absolute ratios.
2. The necessity for international comparison requires the use of *standards* (or *reference materials*) to which the samples have to be related.
3. The use of isotope ratios would lead to reporting results as numbers comprising 5 to 7 digits, these being impractical and hard to remember.
4. Absolute ratios are less relevant than the changes in ratios occurring during certain processes; that is, transitions between phases or molecules.

Therefore, an isotope abundance is generally reported as a deviation of the isotope ratio of a sample A *relative* to that of a reference sample or standard, r:

$$\delta_{A/r} = \frac{R_A}{R_r} - 1 \quad (\times 10^3\text{‰}) \tag{2.37}$$

For the specific isotopes mentioned in Eq. 2.38 the respective δ values are indicated by the symbols:

$$^2\delta = \frac{(^2H/^1H)_A}{(^2H/^1H)_r} - 1 \qquad ^{13}\delta = \frac{(^{13}C/^{12}C)_A}{(^{13}C/^{12}C)_r} - 1 \qquad ^{18}\delta = \frac{(^{18}O/^{16}O)_A}{(^{18}O/^{16}O)_r} - 1$$

$$\tag{2.38}$$

where the more often used symbols are δ^2H or δD, $\delta^{13}C$ and $\delta^{18}O$, respectively.

However, in this volume we will write the mass numbers indicating the rare isotopes concerned by superscripts before the δ's as with the R symbols and the isotope symbols themselves, thus making the symbols simpler and leaving space for other super- and subscripts.

> As the values of δ in nature are small, they are generally given – although *not defined* – in ‰ (per mill ≡ parts per thousand ≡ 10^{-3}). We have to emphasise that ‰ is equivalent to the factor 10^{-3} and is not a unit as is often suggested. *A δ value is merely a small, dimensionless number.* Similarly to ε defined in Section 2.3.1, it is incorrect to use $\delta/10^3$ instead of δ in mathematical equations. In order to avoid neglecting to realise that δ is a small number, one may choose to write in mathematical equations: $\delta(\text{in ‰})10^{-3}$ instead of merely δ, impractical though it is.
> Negative δ values indicate lower abundances of the rare isotope in these samples than in the reference material; positive δ values point to higher abundances.

Comparing the definitions of ε (Eq. 2.6) and δ, the only difference is that whereas ε was defined as the isotope ratio of a compound (then B) relative to another compound (A), δ is defined as the isotope ratio of a compound (now A) relative to the standardised reference material (r).

Although in the equations derived in Chapter 3 the δ format can be applied in practically all natural processes to the precision attainable by modern analytical instrumentation, the calculation capacity of small pocket calculators used in the field has generally made the approximations superfluous. The δ values delivered by isotope laboratories can easily be transformed into R values by applying (rewritten from Eq. 2.37):

$$R = R_r(1 + \delta) \tag{2.39}$$

where the R_r values are the isotope ratios of the international reference materials to be defined later (Chapter 5). Referring to the relation between isotope ratio and isotope concentration (Section 2.1), the isotope concentration, C, can now be transformed into δ by inserting Eq. 2.4:

$$\delta = \frac{R}{R_r} - 1 = \frac{1}{R_r} \frac{C}{1 - C} - 1 \tag{2.40}$$

Examples: natural carbon contains about 1 atom % of ^{13}C; carbon with an absurdly high ^{13}C concentration of 99 atom % thus has a δ value of $8809 \times 10^3‰$ relative to PDB-CO_2 (Pee Dee Belemnite) with $R_r = 0.0112372$, Section 5.1.3).

If δ values are to be given relative to a secondary reference, r', rather than to the primary reference r, a conversion equation is needed.

From the isotope ratios:

$$\frac{R_A}{R_{r'}} = \frac{R_A/R_r}{R_{r'}/R_r} \tag{2.41}$$

or

$$1 + \delta_{A/r'} = \frac{1 + \delta_{A/r}}{1 + \delta_{r'/r}}$$

follows

$$\delta_{A/r'} = \frac{\delta_{A/r} - \delta_{r'/r}}{1 + \delta_{r'/r}} \tag{2.42}$$

This equation is also useful for converting between two phases or compounds A and B:

$$\delta_{A/B} = \frac{-\delta_{B/A}}{1 + \delta_{B/A}} \approx -\delta_{B/A} \tag{2.43}$$

and is similar to Eq. 2.7 for ε. Values of R and α are multiplicative, whereas δ and ε values are only approximately additive. In other words, the fractionation ε, as defined by Eq. 2.6, is almost equal to the difference in isotopic composition between two phases:

$$\varepsilon_{B/A} = \alpha_{B/A} - 1 = \frac{R_B}{R_A} - 1 = \frac{R_B/R_r}{R_A/R_r} - 1 = \frac{1 + \delta_{B/r}}{1 + \delta_{A/r}} - 1 \approx \delta_{B/r} - \delta_{A/r} \tag{2.44}$$

For the reaction A $\rightarrow$ B, $\varepsilon_{B/A}$ is the change in isotopic composition (equal the 'new' δ minus the 'old' (cf. Section 2.3).

Because strict conventions govern the use of references, we will generally drop the subscript r and designate the sample by a subscript δ_A or $\delta(A)$ instead of $\delta_{A/r}$ and $\delta_r(A)$.

The additivity of δ and ε values is commonly approximated by:

$$\delta_B \approx \delta_A + \varepsilon_{B/A} \tag{2.45}$$

This means that if air-CO_2 with a $^{13}\delta$ value of $-8‰$ is assimilated by plants with a carbon isotope fractionation of $-17‰$, the $^{13}\delta$ value of the plant carbon is $-25‰$.

2.9   TRACER CONCENTRATION, AMOUNT OF TRACER

Natural isotopes can be used as *tracers* to follow elements of the water cycle in their natural course. For example, the abundance of ^2H and ^{18}O in precipitation can be applied to unravelling the various components of the runoff hydrograph. This application thus calculates the amount of tracer in the different components of the local or regional water budget.

In Section 2.1 we defined the isotope concentration of ^{18}O in water as:

$$[H_2{}^{18}O] = \frac{\text{amount of } H_2{}^{18}O}{\text{amount of water}} = \frac{H_2{}^{18}O}{H_2{}^{16}O + H_2{}^{18}O} = \frac{{}^{18}R(H_2O)}{1 + {}^{18}R(H_2O)} \tag{2.46}$$
$$\approx {}^{18}R(H_2O)$$

and in the δ notation as:

$$[H_2{}^{18}O] = {}^{18}R_r\{{}^{18}R(H_2O)/{}^{18}R_r\} = {}^{18}R_r\{1 + {}^{18}\delta\} \tag{2.47}$$

However, the rare isotope, as such, cannot simply serve as the tracer. This becomes clear if we consider the case of *runoff hydrograph analysis*. If during a storm rain would fall with a $^{18}\delta$ value equal to that of the *base flow* (=groundwater) there would be no isotopic change in the runoff, and so we would not be able to recognise the precipitation in the runoff. In other words, there would be no recognisable tracer. The precipitation can only be 'traced' if it can be distinguished isotopically from the base flow. Therefore, the *tracer concentration* has to be defined as the deviation of the rare isotope concentration from a certain base level, in our case the $^{18}\delta$ value of the base flow (= average groundwater). The concentration of the tracer is then defined as:

$$[H_2{}^{18}O] - [H_2{}^{18}O_{base}] = {}^{18}R_r\{{}^{18}\delta - {}^{18}\delta_{base}\} \tag{2.48}$$

Here we have inserted $^{18}R_r$ as the ^{18}R value of the international reference (i.e. standard, to be introduced in Chapter 5).

If a comparison is required of, as in our present case, the amount of runoff and the amount of precipitation, we have to introduce the *tracer amount*, defined as:

$$^{18}R_r\{{}^{18}\delta - {}^{18}\delta_{base}\}W = {}^{18}R_r\Delta{}^{18}\delta W \tag{2.49}$$

where W denotes the amount of water. Specific comparison of the amount of runoff Q and the amount of precipitation P now reduces to calculating the ratio:

$$\frac{\sum_i (\Delta^{18}\delta_Q Q)_i}{\sum_j (\Delta^{18}\delta_P P)_j} \tag{2.50}$$

for a number of time periods i and j, respectively (i may equal j, depending on the question).

CHAPTER 3

Isotope fractionation processes

In the preceding chapters we have defined isotope (abundance) ratios in various chemical compounds. We have also discussed the phenomenon of isotope fractionation during some single chemical or physical processes, and presented some qualitative conclusions for the magnitude of isotope fractionation. Here we discuss the calculation of isotopic changes by some more complicated natural processes, once the fractionation of a single process is known.

The derived equations can be expressed almost without approximation as isotope ratios (R values), as well as in an approximated format in δ values.

3.1 MIXING OF RESERVOIRS WITH DIFFERENT ISOTOPIC COMPOSITION

3.1.1 Mixing of reservoirs of the same compound

If two (or more) quantities of a certain compound with different δ values are being mixed (e.g. CO_2 in the air of two different origins, such as marine and biospheric), the δ value of the mixture can be calculated from a *mass balance* equation (Fig. 3.1A).

N_1 molecules of a certain compound with an isotope ratio R_1 contain $N_1/(1 + R_1)$ molecules of the abundant, and $R_1N_1/(1 + R_1)$ molecules of the rare isotopic species (cf. Eq. 2.3). If this quantity is mixed with another quantity N_2 (with isotopic composition R_2),

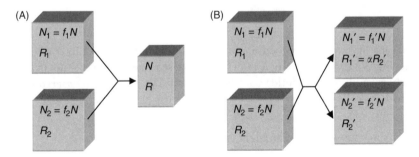

Figure 3.1. Schematic representation of (A): the mixing of two quantities of one compound with different isotopic compositions and (B): the mixing and reaction of two different compounds or two phases of the same compound with different isotopic composition that will be subjected to isotope exchange and will ultimately become a fractionation factor (α) apart.

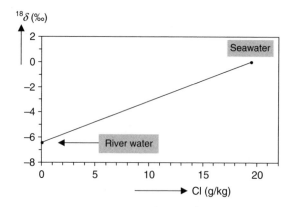

Figure 3.2.   Linear relation between the chlorinity (Cl concentration) and $^{18}\delta$ of water in the estuary of the Western Scheldt, where water from the Belgian River Scheldt (Cl = 0‰ = grams of chloride per kg of seawater; $^{18}\delta = -6.5‰$) mixes with water from the North Sea (Cl = 19.36‰; $^{18}\delta = 0‰$).

a simple mass balance gives the isotope ratio for the mixture:

$$\frac{RN}{1+R} = \frac{R_1 N_1}{1+R_1} + \frac{R_2 N_2}{1+R_2} \quad \text{or} \quad RN \approx R_1 N_1 + R_2 N_2 \tag{3.1}$$

where $N = N_1 + N_2$. The approximation is valid for $R \approx R_1 \approx R_2$ (note that the R values have not been neglected with respect to 1, but rather with respect to each other).

For natural abundances of isotopes within a δ range of 100‰, the R values can be approximated (to within $\pm 0.03‰$) by $R_r(1 + \delta)$ (Eq. 2.39) resulting in:

$$(1 + \delta)N = (1 + \delta_1)N_1 + (1 + \delta_2)N_2$$

and with $N = N_1 + N_2$:

$$\delta N = \delta_1 N_1 + \delta_2 N_2 \tag{3.2}$$

Denoting the fractional contributions to the mixture by $f_1 \; (= N_1/N)$ and $f_2 \; (= N_2/N)$, where $f_1 + f_2 = 1$, the δ value of the mixture is:

$$\delta = f_1 \delta_1 + f_2 \delta_2 \tag{3.3}$$

which means that, to a very good approximation, additivity applies to δ values. As an example, Fig. 3.2 shows the isotopic effect of mixing seawater and river water in an estuary. Additivity is valid for the chloride content (or salinity) of the water (*conservative tracer*) as well as for $^{18}\delta$. The relation between the two is consequently a straight line.

3.1.2   Mixing of reservoirs of different compounds

If the mixing process includes chemically different but reactive compounds of quantities N_1 and N_2 with different isotopic compositions R_1 and R_2, the final stage is characterised by a fractionation between the compounds. During the reaction a chemical shift between the two compounds will often occur. The quantities change into N_1' and N_2' with isotopic compositions of R'_1 and R'_2, where the latter values are a fractionation factor α apart. An example is the exchange between gaseous CO_2 and a bicarbonate solution such as between lake water and atmospheric CO_2, or the mixing of freshwater and seawater with different pH values.

A similar mass balance for the rare isotope can be written (Fig. 3.1B):

$$\frac{R'_1 N'_1}{1+R'_1} + \frac{R'_2 N'_2}{1+R'_2} = \frac{R_1 N_1}{1+R_1} + \frac{R_2 N_2}{1+R_2}$$

or

$$R'_1 N'_1 + R'_2 N'_2 \approx R_1 N_1 + R_2 N_2 \tag{3.4}$$

where $N'_1 + N'_2 = N_1 + N_2$, and the same approximation applies as in Eq. 3.1 for the range of δ values of all natural samples.

This results in:

$$R'_2 = \frac{R_1 N_1 + R_2 N_2}{\alpha N'_1 + N'_2} \quad \text{and} \quad R'_1 = \alpha R'_2 \tag{3.5}$$

or

$$\delta'_2 \approx \frac{f_1 \delta_1 + f_2 \delta_2 - f'_1 \varepsilon}{1 + f'_1 \varepsilon} \quad \text{and} \quad \delta'_1 \approx \delta'_2 + \varepsilon \tag{3.6}$$

with $f_1 = N_1/(N_1 + N_2)$, etc.

These equations are applied to the indirect determination of $^{18}\delta$ of a water sample, viz. by measuring $^{18}\delta$ of CO_2 equilibrated with the water rather than the water itself (Appendix I, Section I.2.1).

3.2   ISOTOPIC CHANGES IN RAYLEIGH PROCESSES

It is common in isotope studies to be concerned with the change in isotopic composition of a reservoir because of the removal of an increasing fraction of its contents accompanied by isotope fractionation. We will discuss the process under two conditions: (i) removal of a compound ($= sink$) from a reservoir without an additional input (e.g. evaporation from a confined water basin); (ii) removal accompanied by a constant input ($= source$) (such as evaporation from a lake with a river input).

3.2.1   Reservoir with one sink

A simple box model describing a reservoir with only one sink is presented by Fig. 3.3. Here N is the total number of molecules, and R is the ratio of the rare to abundant molecular concentrations, so that

$N/(1+R)$ is the number of abundant isotopic molecules and

$RN/(1+R)$ is the number of rare isotopic molecules

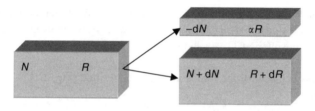

Figure 3.3.   Schematic representation of the Rayleigh process. N is the number of abundant isotopic molecules ($\approx$ total number of molecules), R is the isotope ratio, α is the (equilibrium or kinetic) fractionation factor involved.

For the removal of dN molecules with an accompanying fractionation factor α, the mass balance for the rare isotope gives:

$$\frac{R}{1+R}N = \frac{R+dR}{1+R+dR}(N+dN) - \frac{\alpha R}{1+\alpha R}dN$$

To an acceptable approximation we will take the total number of molecules equal to the number of abundant isotopic molecules; in fact, for this approximation all denominators are taken equal to $1+R$ (here R is not neglected with respect to 1). The mass balance for the rare isotope now becomes:

$$RN = (R+dR)(N+dN) - \alpha R dN$$

and neglecting the products of differentials:

$$dR/R = (\alpha - 1)dN/N$$

Integration and applying the boundary condition $R = R_0$ at $N = N_0$ gives:

$$\frac{R}{R_0} = \left(\frac{N}{N_0}\right)^{\alpha-1} \tag{3.7}$$

or in δ values with respect to the standardised reference:

$$\delta = (1 + \delta_0)(N/N_0)^{\varepsilon} - 1 \tag{3.8}$$

where N/N_0 represents the remaining fraction of the original reservoir; R_0 and δ_0 refer to the original isotopic composition. Again we remind the reader that $\varepsilon (= \alpha - 1)$ is a small number, for instance -0.004 if the fractionation is $-4\%_0$.

The increase in $^{18}O/^{16}O$ of a confined water body by evaporation with $\alpha < 1$ (Fig. 3.4 upper curve) serves as an example of this simple Rayleigh process. Another case is presented by the $^{13}C/^{12}C$ increase in lake-water bicarbonate due to the growth of plankton.

We have until now focused our attention on the isotopic change of the reservoir. It is also relevant to calculate the isotopic composition of the compound which is gradually built up by the infinitesimal increments dN. This is not a simple additive procedure, because the isotopic composition of the reservoir changes during the process and thus also that of the increments dN. An example is found in the formation of a carbonate deposit from a groundwater mass which loses CO_2 when exposed to the atmosphere.

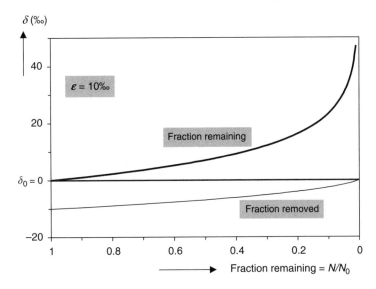

Figure 3.4.   The enrichment of a rare heavy isotope in a reservoir by the Rayleigh process as a function of the fraction removed. The preferential removal of the light isotopic species is accompanied by a fractionation of $\alpha < 1$. The lower line presents the average cumulative isotopic composition of the compound formed by the process, δ_Σ (Eq. 3.10).

The average isotopic composition of the formed compound ($=$ the sum (Σ) of the increments) at a certain time is:

$$R_\Sigma = \frac{1}{N_0 - N} \int_N^{N_0} \alpha R dN = \frac{1}{N_0 - N} \frac{\alpha R_0}{N^{\alpha-1}} \int_N^{N_0} N^{\alpha-1} \, dN$$

where R is given by Eq. 3.7. Integration of this equation gives:

$$R_\Sigma = \frac{N_0}{N_0 - N} R_0 \left(\frac{N}{N_0}\right)^\alpha \Bigg|_N^{N_0} = R_0 \frac{1 - (N/N_0)^\alpha}{1 - N/N_0} \tag{3.9}$$

or

$$\delta_\Sigma = (1 + \delta_0) \frac{1 - (N/N_0)^\alpha}{1 - N/N_0} - 1 \tag{3.10}$$

The resulting change in δ_Σ (for $\delta_0 = 0$) is shown in Fig. 3.4 by the lower curve.

3.2.2   Reservoir with two sinks

A slightly more complicated, but nevertheless realistic phenomenon occurs if two processes simultaneously withdraw two different compounds from the reservoir (Fig. 3.5). This process occurs when, through growth of algae in surface water, the pH of the water increases causing the formation of calcium carbonate. Both processes involve characteristic carbon isotope fractionation (see Chapter 5).

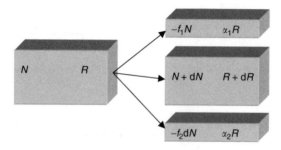

Figure 3.5.   Schematic representation of the Rayleigh process where two compounds are being removed from the reservoir with different fractionation (α_1 and α_2); f_1 and f_2 refer to the fractional contributions to the total sink ($f_1 + f_2 = 1$).

The rare isotope mass balance now gives:

$$RN = (R + dR)(N + dN) - \alpha_1 R f_1 dN - \alpha_2 R f_2 dN$$

Here $f_1 + f_2 = 1$; a total amount dN of the element under consideration is removed from the reservoir. As with the previous case, we have:

$$\frac{dR}{R} = (\alpha_1 f_1 + \alpha_2 f_2 - 1) \frac{dN}{N}$$

which after integration gives:

$$R = R_0 \left(\frac{N}{N_0} \right)^{\alpha_1 f_1 + \alpha_2 f_2 - 1} \tag{3.11}$$

or

$$\delta = (1 + \delta_0)(N/N_0)^{\varepsilon_1 f_1 + \varepsilon_2 f_2} \tag{3.12}$$

This reduces to the simple case of one sink (Eqs 3.7 and 3.8) if $f_2 = 0$, while the exponent reduces to $\varepsilon_1 f_1$ if the removal of the second compound occurs without fractionation.

3.2.3   Reservoir with one source and one sink, as a function of time

Although the previous are practical examples of Rayleigh processes, the general case includes the existence of a source as well as a sink. These are represented by a supply rate i (in quantity per unit time) and a loss rate u (Fig. 3.6A), while the fractionation with input, α_i, is not necessarily equal to unity.

The rare isotope mass balance is:

$$RN + \alpha_i R_i i dt = (R + dR)[N + (i - u)dt] + \alpha R u dt$$

from which we have by separation of variables (with $\alpha - 1 = \varepsilon$):

$$\frac{dR}{(i + \varepsilon u)R - \alpha_i R_i i} = \frac{dt}{N_0 + (i - u)(t - t_0)}$$

where N at a time t has been replaced by the denominator on the right-hand side ($N = N_0$ at $t = t_0$).

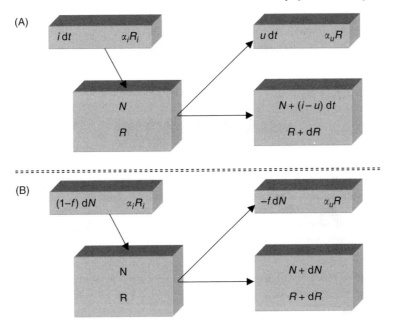

Figure 3.6.   Schematic representation of the Rayleigh process with one source and one sink: (A) as a function of time with i and u denoting the input and output rates respectively; and (B) as a function of the quantities of inflow and outflow with f denoting the ratio between input and output rates.

Integration, and applying the boundary condition that $R = R_0$ at $t = t_0$, results in:

$$R = \frac{\alpha_i R_i}{1 + (u/i)\varepsilon} + \left[R_0 - \frac{\alpha_i R_i}{1 + (u/i)\varepsilon}\right]\left[1 + \frac{i - u}{N_0}(t - t_0)\right]^{-(i+u\varepsilon)/(i-u)} \qquad (3.13)$$

for the case where $i \neq u$.

If $i = u$, the mass balance is:

$$RN_0 + \alpha_i R_i i \, dt = (R + dR)N_0 + \alpha R u \, dt$$

from which we can similarly derive:

$$R = \frac{\alpha_i R_i}{\alpha} + \left[R_0 - \frac{\alpha_i R_i}{\alpha}\right]e^{-\alpha(i/N_0)(t-t_0)} \qquad (3.14)$$

As $t \to \infty$, a steady-state situation is reached with:

$$R = \frac{\alpha_i R_i}{\alpha} \quad \text{or} \quad \delta = \frac{(1 + \varepsilon_i)(1 + \delta_i)}{1 + \varepsilon} \qquad (3.15)$$

which is approximated by

$$\delta = \delta_i + (\varepsilon_i - \varepsilon) \qquad (3.16)$$

Such a situation occurs where the evaporation of a lake is balanced by the inflow of water from a river (Dead Sea, Lake Titicaca, Lake Chad). In this case $\alpha_i = 1$ because the inflow occurs without fractionation.

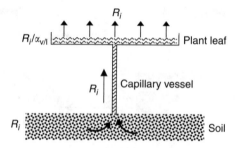

Figure 3.7.   Schematic representation of the evapotranspiration process. The capillary flow of the soil water with isotope ratio R_i (for ^{18}O or ^2H) occurs without fractionation. At the surface of the plant leaf evaporation is accompanied by fractionation $\alpha_{v/l}$ (Table 5.4 or 5.6). In a stationary state the escaping vapour should have the same R value as the transpired water. The result is that the (supposedly well-mixed) leaf water is enriched in ^{18}O and ^2H by $\alpha_{v/l}$.

The meaning and value of α have to be considered separately, as α describes the incongruent dissolution process which occurs in some groundwater systems. The result is simply understood by noting that a steady-state requires the input to be isotopically equal to the output ($R_i = \alpha R$).

Another important example is the fact that *evapotranspiration* (the phenomenon that water from the soil is taken up by a plant through capillary action of the plant vessels, and subsequently evaporates from the leaf surface) causes water to be transported by a plant from the soil to the air with no net isotope fractionation. The process is demonstrated by Fig. 3.7.

The small diameter of the capillary results in a relatively high flow rate. This is crucial, because it prevents the isotopically heavy leaf water to diffuse downward.

If there is no source, $i = 0$ and Eq. 3.13 becomes similar to Eq. 3.7, the simple case with only one sink.

If the sink is absent, the isotopic composition approaches $R = \alpha_i R_i$ for $t \to \infty$.

In general cases where $i \neq u$ (e.g. slowly increasing or decreasing surface water bodies, when $\alpha_i = 1$), the resulting isotopic composition of the reservoir depends on competition between the sink and source. If $u < i$ and for $t \to \infty$, the value of R approaches $\alpha_i R_i[1 + (u/i)\varepsilon]$, while in the opposite case ($u > i$) $R \to \infty$, as with the simple Rayleigh process discussed in Section 3.2.1.

3.2.4   Reservoir with one source and one sink, as a function of mass

If the Rayleigh process involves one source and one sink there is an alternative mathematical approach in which the isotopic composition is considered to be a function of changes in quantity N, rather than of time (Fig. 3.6B). The isotope mass balance now gives:

$$RN + \alpha_i R_i(1 - f)dN = (R + dR)(N + dN) - \alpha Rf dN$$

This results in the differential equation:

$$\frac{dR}{(\alpha f - 1)R + (1 - f)\alpha_i R_i} = \frac{dN}{N}$$

Integration and application of the boundary condition that $R = R_0$ at $N = N_0$ results in:

$$R = \frac{(1-f)\alpha_i}{1-\alpha f}R_i + \left[R_0 - \frac{(1-f)\alpha_i}{1-\alpha f}R_i\right]\left(\frac{N}{N_0}\right)^{\alpha f - 1} \tag{3.17}$$

where N/N_0 denotes the remaining fraction of the original reservoir. A common example of this process in nature is represented by the evaporation of an isolated water body. With decreasing size of the reservoir N, R approaches:

$$R_\infty = \frac{(1-f)\alpha_i}{1-\alpha f} \tag{3.18}$$

3.2.5   Reservoir with two sources and two sinks, with and without fractionation

Finally, we treat the conditions at a stationary state; that is, a situation in which inflow and outflow of a well-mixed reservoir are balanced, such that the isotopic composition of the reservoir compound has become constant over time (Fig. 3.8). A practical example of this model is presented by the infinitesimal part of a river at a certain location in a tropical region where evaporation is significant, resulting in gradually increasing $^2\delta$ and $^{18}\delta$ values along the course of the river; α represents the evaporation and exchange with atmospheric vapour; I_0 is the original water flow rate at the river source.

The stationary state requires that the sinks and sources of the infinitesimal part are balanced for the abundant as well as for the rare isotope. This condition is mathematically presented by:

$$I + dI + \frac{f}{1-f}dI = \frac{1}{1-f}dI + I$$

and

$$(R + dR)(I + dI) + \alpha_i R_i \frac{f}{1-f}dI = \alpha_u R \frac{1}{1-f}dI + RI$$

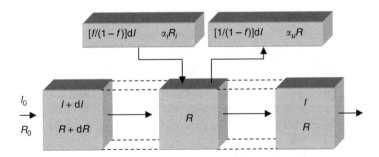

Figure 3.8.   Model representing a well-mixed reservoir with simple inflow and outflow, as well as with a source and sink affecting the isotopic composition of the reservoir. For example, during an infinitesimal time period dI has escaped from the (river) water into the air through evaporation and exchange with the ambient vapour.

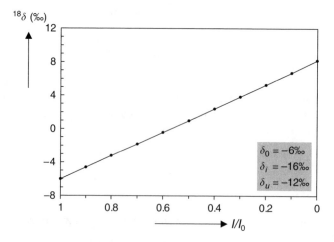

Figure 3.9.   Result of the model of Fig. 3.8, representing a river which evaporates continuously during its course (Eq. 3.20). The value of f has been chosen as 0.5. The δ and ε values are realistic. The shape of the curve is almost a straight line, contrary to the simple evaporation presented in Fig. 3.4. A similar curve can be calculated for $^2\delta$. The ratio between these two curves is discussed in Section 5.5.

or

$$\left(\frac{\varepsilon_u + f}{1-f}R - \frac{\alpha_i f}{1-f}R_i\right)dI = IdR$$

Integration and applying the boundary condition $R = R_0$ at $I = I_0$ results in:

$$\ln\frac{\{[(\varepsilon_u + f)/(1-f)]R - [(\alpha_i f)/(1-f)]R_i\}}{\{[(\varepsilon_u + f)/(1-f)]R_0 - [(\alpha_i f)/(1-f)]R_i\}} = \frac{\varepsilon_u + f}{1-f}\ln\frac{I}{I_0}$$

Finally, after some rearrangement:

$$R = \frac{\alpha_i f}{\varepsilon_u + f}R_i + \left[R_0 - \frac{\alpha_i f}{\varepsilon_u + f}R_i\right]\left(\frac{I}{I_0}\right)^{(\varepsilon_u + f)/(1-f)} \tag{3.19}$$

or

$$\delta = \frac{\alpha_i f}{\varepsilon_u + f}(1+\delta_i) + \left[1 + \delta_0 - \frac{\alpha_i f}{\varepsilon_u + f}(1+\delta_i)\right]\left(\frac{I}{I_0}\right)^{(\varepsilon_u + f)/(1-f)} - 1 \tag{3.20}$$

The result of the last model calculation is shown in Fig. 3.9 for δ and ε values as indicated in the graph. In the case of flowing river water that is subjected to evaporation, the value of f depends on the relative humidity. For the calculation of Fig. 3.9, f has arbitrarily been chosen as 0.5.

The difference from the simple, one-directional evaporation of Fig. 3.3 is obvious; while in the former case δ of the remaining water approaches infinity, here the δ increase is very regular. The important result of comparing the trends for $^{18}\delta$ and $^2\delta$ with increasing I/I_0 are discussed in Section 5.5 and 8.1.

CHAPTER 4

Radioactive decay and production

This section contains a brief review of basic facts about radioactivity, that is, the phenomenon of *nuclear decay* in combination with (*radioactive*) *radiation*, and about nuclear reactions, the production of nuclides. For full details the reader is referred to textbooks on nuclear physics and radiochemistry, and on the application of radioactivity in the earth sciences.

4.1 NUCLEAR INSTABILITY

If a nucleus contains too great an excess of neutrons or protons, it will sooner or later disintegrate to form a stable nucleus. The different modes of radioactive decay will be discussed briefly; the various changes that can take place in the nucleus are shown in Fig. 4.1. In transforming into a more stable nucleus the unstable nucleus loses potential energy, part of its binding energy. This energy Q is emitted as kinetic energy and is shared by the particles formed according to the laws of conservation of energy and momentum (see Section 4.2.8).

The energy released during nuclear decay can be calculated from the mass budget of the decay reaction, using the equivalence of mass and energy discussed in Section 1.4. The atomic mass determinations have been made very accurately and precisely by mass spectrometric measurement:

$$E = [M_{\text{parent}} - M_{\text{daughter}}] \times mc^2 \tag{4.1}$$

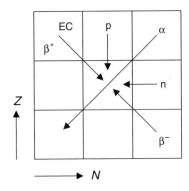

Figure 4.1.   Different modes of radioactive decay visualised on one spot of the Chart of Nuclides (cf. Fig. 1.2). N is the number of neutrons, Z the number of protons.

with

$$1 \text{ amu} \equiv 931.5 \text{ MeV} \tag{4.2}$$

Disintegration of a parent nucleus produces a daughter nucleus which generally is in an *excited state*. After an extremely short time the daughter nucleus loses this excitation energy through the emittance of one or more γ (*gamma*) *rays*, electromagnetic radiation with a very short wavelength similar to the emission of light (also electromagnetic radiation, but with a longer wavelength) by atoms which have been brought into an excited state. In some cases, however, β decay is directly to the ground state of the daughter nucleus, without the emission of γ radiation. Fig. 4.2 shows the various *decay schemes* for some familiar nuclides.

4.2   NUCLEAR DECAY AND RADIATION

In the following sections the various types of spontaneous nuclear disintegration are discussed.

4.2.1   Negatron (β⁻) decay

The most abundant decay mode is β⁻ decay. This involves the emission of a negative electron, a *negatron* or β⁻ (*beta*) *particle* because of a transformation inside the nucleus of a neutron into a proton:

$$n \rightarrow p^+ + \beta^- + \bar{\nu} + Q \tag{4.3}$$

This results in a daughter nucleus with equal A and a change in atomic number Z to $Z + 1$, and the emission of an electron and an anti-neutrino. An (anti-)neutrino is a particle with mainly a relativistic mass, that is, *mass because of its motion* (neutrinos and negatrons have their spin anti-parallel to their direction of motion, and anti-neutrino's and positrons parallel).

The total reaction energy Q is as kinetic energy shared between the β particle and the (anti-)neutrino. The consequence is that, although Q is a specific amount of energy, the energy spectrum of the β⁻ particles is continuous from zero to a certain maximum energy at $E_{\beta \max} = Q$. The energy distribution is such that the maximum in this distribution is at an energy about one third of the maximum energy.

As mentioned earlier, in some cases the resulting daughter nucleus is not in an excited state, so that no γ radiation is emitted. This happens to be the case with the two radioactive isotopes that are most relevant in isotope hydrology, namely ^{3}H and ^{14}C.

4.2.2   Positron (β⁺) decay

This type of nuclear transformation involves the emission of a *positron* or β⁺ (*beta plus*) *particle* as the result of a transformation inside the nucleus of a proton into a neutron:

$$p^+ \rightarrow n + e^+ + \nu + Q \tag{4.4}$$

and, because after slowing down the positron reacts rapidly with any electron present:

$$e^+ + e^- \rightarrow 2\gamma \tag{4.5}$$

Here two electron masses are transformed into energy. The process is called *annihilation*, and the energy released *annihilation energy*. With the mass of the electron

$$m_e = 5.4858026 \times 10^{-4} \text{ amu} \tag{4.6}$$

and using Eq. 4.2, the resulting energy is 1.022 MeV, which is represented by two oppositely directed γ particles of 511 keV each. This type of decay is, for example, known from naturally occurring (because long-living) ^{40}K (Fig. 4.2).

4.2.3 Electron capture (EC)

The nucleus may pick up one electron from the atomic electrons circling the nucleus in consecutive shells. In that case the following reaction takes place:

$$p^+ + e^- \rightarrow n + v + Q \tag{4.7}$$

Depending on from which atomic shell the electron is caught by the nucleus, electron capture is called K capture, L capture and so on. β^+ decay and EC leave the mass number A of the nucleus unchanged, while Z becomes $Z - 1$ (Fig. 4.1). The atom is left in an excited state and returns to the ground state by emitting short-wave electromagnetic radiation, namely, low-energy γ rays or X-rays. As in the previous decay modes, the excited state of the nucleus causes the emission of one or more γ rays.

4.2.4 Alpha (α) decay

α (alpha) decay involves the emission of an α *particle* (a ^{4}He nucleus):

$$^A_Z X \rightarrow ^{A-4}_{Z-2} Y + ^4_2 He \tag{4.8}$$

This type of decay occurs primarily within heavy elements such as in the uranium and thorium series. The emission of the heavy α particle coincides with the emission of relatively large-energy γ rays.

Because, as in all cases, the reaction energy Q is shared by the atoms Y and He according to the two conservation laws, the case of α emission is special. As with a gun shot, where the rifleman feels a large momentum against his shoulder, the daughter nucleus receives a relatively large so-called *recoil* (Section 4.2.8). This is particularly important when the *recoil energy* is larger than the molecular binding energy and sufficiently large to loosen the daughter nucleus from its chemical bond. This phenomenon is the basis for studying the natural abundance of the decay nuclides of uranium and thorium.

4.2.5 Branching decay

Radioactive nuclides that show two different modes of decay also exist in nature. One example is presented by ^{40}K, that can decay with the emission of a β^- or a β^+ particle (Fig. 4.2).

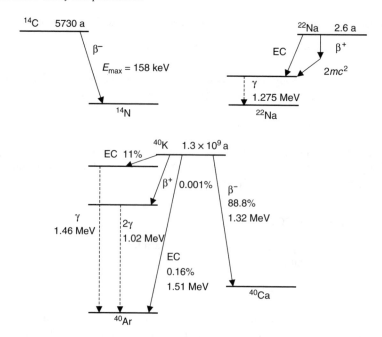

Figure 4.2.   Decay schemes for some nuclides obeying β^-, β^+ or EC decay, the latter example of ^{40}K in branching decay and the consecutive γ radiation by the excited daughter nucleus as well as the formation of 2γ by annihilation of an emitted β^+ particle.

Each decay mode has its specific decay constant or half-life (see Section 4.3). The total decay is simply the sum of both single chances and is thus given by:

$$\lambda_{\text{total}} = \lambda_1 + \lambda_2 \tag{4.9}$$

and the total half-life consequently by:

$$(1/T_{1/2})_{\text{total}} = (1/T_{1/2})_1 + (1/T_{1/2})_2 \tag{4.10}$$

4.2.6   Spontaneous and induced fission, neutron emission

Nuclides of the heavy elements have the possibility of breaking into two relatively large daughter nuclei, together with the emission of a small number of neutrons. This decay may occur spontaneously as well as induced by the bombardment of the parent nucleus by a neutron from elsewhere, for instance from a neighbouring nucleus. This is the dominant process in nuclear reactors (Sections 4.4.2).

4.2.7   Radioactive decay series

If a radioactive nuclide is situated in the Chart of Nuclides (Section 1.2) far from the stability line (for the light elements at $Z = N$), the daughter nucleus after radioactive decay may be radioactive as well. In nature this occurs with the heavy nuclides in the uranium and thorium decay series. Here the original decay of ^{238}U, ^{235}U and ^{232}Th is followed by a series of

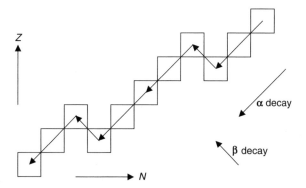

Figure 4.3.   Schematic representation of a hypothetical multiple decay curve, analogous to the decay curves of the U and Th decay series.

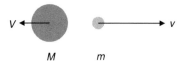

Figure 4.4.   After radioactive decay the daughter nucleus M and the smaller emitted particle m (for instance an α particle) move in opposite direction, together carrying the decay energy as the respective kinetic energies.

radioactive decay products, finally ending with a stable isotope of lead (Pb). Fig. 4.3 shows schematically how the elements of such decay series are related.

We will not go into detail on this as the applications, including elements of the decay series, are mainly outside the scope of Isotope Hydrology. The reader is directed to the complete volumes of the UNESCO/IAEA series.

4.2.8   Recoil by radioactive decay

In the preceding chapter we mentioned the phenomenon of the recoil momentum and energy which any decaying nucleus experiences. In the case of a heavy particle such as the α particle leaving the nucleus, the effect is not small. We will now calculate the recoil energy by considering the classical mechanics of this process (Fig. 4.4).

The *principle of conservation of momentum* requires:

$$MV + mv = 0 \quad \text{or} \quad M^2V^2 = m^2v^2 \tag{4.11}$$

while the *principle of conservation of energy* requires the decay energy to be equal to the sum of the two kinetic energies:

$$E_{\text{decay}} = \frac{1}{2}MV^2 + \frac{1}{2}mv^2 = E_M + E_m \tag{4.12}$$

From Eq. 4.11 we have:

$$E_M = \frac{1}{2}MV^2 = \frac{1}{2M}M^2V^2 = \frac{1}{2M}m^2v^2 = \frac{m}{M}E_m \tag{4.13}$$

With Eq. 4.12:

$$E_{decay} = \left(1 + \frac{m}{M}\right) E_m$$

so that the two particle energies are, respectively:

$$E_m = \frac{M}{M+m} E_{decay} \quad \text{and} \quad E_M = \frac{m}{M+m} E_{decay} \tag{4.14}$$

In the case of the α decay of ^{238}U with a decay energy of about 4.2 MeV, the daughter nucleus, ^{234}Th, receives a recoil energy of $[4/(234+4)] \times 4.2$ MeV $= 71$ keV, more than enough to break the chemical bond between the daughter (Th) atom and the surrounding atoms in the mineral (see Volumes I and IV of the UNESCO/IAEA series on Groundwater for a discussion of the consequences).

4.3   LAW OF RADIOACTIVE DECAY

4.3.1   Exponential decay

The mathematical expressions presented in this chapter are generally applicable to all those processes in which the transition of the *parent nucleus* to a *daughter nucleus*, that is, the process of radioactive decay, is governed by statistical chance. This chance of decay is equivalent to the degree of instability of the parent nucleus. Each radioactive nuclide has its specific degree of instability which, as we will see, is expressed by the half-life assigned to this nuclide.

The radioactivity of a sample is more complicated if it consists of two or more components, such as:

- in the case of a mixture of independent activities,
- if one specific type of nuclide shows two modes of decay, so-called branching decay and
- if we are dealing with a nuclear decay series in which the daughter nuclides are also radioactive. All these phenomena are discussed separately.

The fundamental law of radioactive decay is based on the fact that the decay, that is, the transition of a parent nucleus to a daughter nucleus, is a purely statistical process. The disintegration (decay) probability is a fundamental property of an atomic nucleus and remains equal in time; *it does not depend on outside influences*.

Mathematically this law is expressed as:

$$dN = \lambda N dt \tag{4.15}$$

and

$$\lambda = \frac{(-dN/dt)}{N} \tag{4.16}$$

where N is the number of radioactive nuclei, $-dN/dt$ the decrease (negative) of this number per unit of time, and λ is thus the probability of decay per nucleus per unit of time. This *decay constant* λ is specific for each decay mode of each nuclide.

The *radioactivity* or *decay rate* is defined as the number of disintegrations per unit of time:

$$A = -dN/dt = \lambda N \tag{4.17}$$

By integration of this relation and applying the boundary condition that in the beginning at $t = 0 \quad N = N^0$, we obtain:

$$\ln(N/N^0) = -\lambda t \tag{4.18}$$

and subsequently the equation of *exponential decay*:

$$N = N^0 e^{-\lambda t} \tag{4.19}$$

or using Eq. 4.17:

$$A = A^0 e^{-\lambda t} \tag{4.20}$$

The time during which A^0 decreased to A (= the age of the material) is:

$$T = -(1/\lambda)\ln(A/A^0) \tag{4.21}$$

The relations of Eqs 4.19 and 4.20 indicate the rate at which the original number of radioactive nuclei (N^0) and the original radioactivity (A^0) decrease in time (Fig. 4.5).

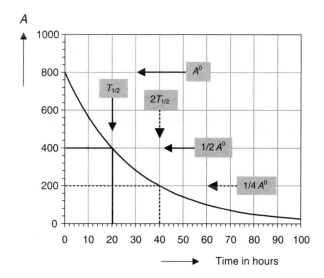

Figure 4.5. The rate of radioactive decay. After each subsequent half-life of 20 hours the number of radioactive nuclei and the original radioactivity of 800 units are halved.

4.3.2   Half-life and mean life

It is a common practice to use the *half-life* ($T_{1/2}$) instead of the decay constant (λ) for indicating the degree of instability or the decay rate of a radioactive nuclide. This is defined as the period of time in which half of the radioactivity has disappeared (half of the nuclei have disintegrated, Fig. 4.5):

$$T_{1/2} = (-1/\lambda)\ln(1/2) \tag{4.22}$$

from which:

$$\lambda = \frac{\ln 2}{T_{1/2}} = \frac{0.693}{T_{1/2}} \tag{4.23}$$

The *mean life* of a nuclide is the sum of the life times of a certain number of nuclei (before they have all disintegrated) divided by the number of nuclei. During the time interval dt a number of dN nuclei disintegrate. These 'lived' during a period t, which amounts to a total life time for dN nuclei of:

$$t\,dN = t\lambda N\,dt$$

Integrating over all nuclei (N) gives the mean life (time):

$$\tau = \frac{1}{N^0} \int_0^\infty t\lambda N\,dt = \lambda \int_0^\infty te^{-\lambda t}\,dt = \lambda \left\{ -\frac{t}{\lambda}e^{-\lambda t}\bigg|_0^\infty + \frac{1}{\lambda}\int_0^\infty e^{-\lambda t}\,dt \right\} \tag{4.24}$$

$$= \lambda \left\{ 0 + \frac{1}{\lambda}\left(-\frac{1}{\lambda}e^{-\lambda t}\bigg|_0^\infty \right) \right\} = \frac{1}{\lambda}$$

As an example, the mean life of a ^{14}C nucleus with $T_{1/2} = 5730$ a is 8267 years. Then $\lambda = 1/8267$, which means that a ^{14}C sample activity decreases by 1‰ (=1/1000) in about 8 years; a ^{3}H sample activity ($T_{1/2} = 12.32$ a) decreases by 5.6% per year.

4.3.3   Activity, specific activity and radionuclide concentration

The *activity* of a certain sample is the number of radioactive disintegrations per second for the sample as a whole. The *specific activity*, on the other hand, is defined as the number of disintegrations *per unit weight* or *volume* of sample (see specific activity of ^{14}C and ^{3}H, Chapter 6). The unit of radioactivity is the *Becquerel* (Bq), which is defined as a decay rate of one disintegration per second (dps), or the now obsolete Curie (Ci) which was defined as a decay rate of 3.7×10^{10} dps.

As an example of the relation between the specific activity of a sample and the actual *concentration* of the radioactive nuclide, we calculate the specific tritium (^{3}H) activity of water containing one ^{3}H atom per 10^{18} hydrogen atoms, by definition equivalent to 1 TU = Tritium Unit (principally wrong name, because *'unit' implies a quantity with dimension*):

$$A_{\text{spec}} = \lambda N \text{(per litre)}$$

where:

$$\lambda = (\ln 2)/T_{1/2} = (\ln 2)/12.32 \text{ a} \quad (1 \text{ year} = 3.16 \times 10^7 \text{ s})$$

$$N = 2 \times 10^{-18} \times (G/M) \times A \quad (G/M = \text{number of moles})$$

$$A = \text{Avogadro's number} = 6.02 \times 10^{23}/\text{mole}$$

$$M = \text{molecular weight} = 18.0$$

$$\text{pCi} = \text{picoCurie} = 10^{-12} \text{ Ci} = 3.7 \times 10^{-2} \text{ dps} = 0.037 \text{ Bq}$$

The numerical result for water with 1 TU of tritium (^{3}H) is:

$$A_{\text{spec}} = 0.119 \text{ Bq/L} = 3.21 \text{ pCi/L} \tag{4.25}$$

As another example we can calculate the ^{14}C concentration in carbon, which has a specific ^{14}C activity of 13.56 dpm per gram of carbon (in the year AD 1950) (see ^{14}C standard activity, Chapter 6):

$$\frac{^{14}\text{C}}{\text{C}} = \frac{13.56 \times 5730 \times (3.16 \times 10^7) \times 12}{60 \times \ln 2 \times (6.02 \times 10^{23})} = 1.2 \times 10^{-12} \tag{4.26}$$

4.3.4   Accumulation of stable daughter product

In some cases we have to pay attention to the amount of 'daughter' produced by the radioactive decay of the 'parent'. In geochemistry an important application is the dating method based on measuring the ^{40}Ar amount accumulated in rock from the decay of ^{40}K (Fig. 4.2). Another example is the accumulation of ^{3}He in water during the decay of ^{3}H. We will take the latter process as an example to conclude the age (T) of a water sample. The principle discussed here is to determine the ^{3}H activity left in the water and compare this to the 'original' ^{3}H activity (cf. Eq. 4.21):

$$T = -(T_{1/2}/\ln 2) \ln(^3A/^3A^0) = -17.8 \ln(^3N/^3N^0) \tag{4.27}$$

where the activity A can be replaced by the amount N. The subscript refers to the original activity/amount/concentration. Instead of measuring the amount of parent remaining, the accumulated amount of daughter product can be taken as representing the age:

$$N(^3\text{He accumulated}) = {}^3N_0(\text{original } {}^3\text{H}) - {}^3N(\text{remaining } {}^3\text{H after time } T)$$

or:

$$N(^3\text{He}) = {}^3N^0(1 - e^{-\lambda T}) = \frac{^3A^0}{\lambda}(1 - e^{-\lambda T}) = \frac{^3Ae^{\lambda T}}{\lambda}(1 - e^{-\lambda T})$$
$$= (^3A/\lambda)(e^{\lambda T} - 1) \tag{4.28}$$

where the amount of gas accumulated (V in litres STP) is related to the number of atoms $N(^3\text{He})$ by:

$$V = \frac{N(^3\text{He})}{6 \times 10^{23}} 22.4 \text{ L}$$

so that:

$$\frac{6 \times 10^{23}}{22.4} V = \frac{A_1}{\lambda} (e^{\lambda T} - 1)$$

and the period of time elapsed since time zero (the 'age' T) is:

$$T = \frac{1}{\lambda} \ln \left(2.7 \times 10^{22} \frac{\lambda}{A_1} V + 1 \right) \tag{4.29}$$

This dating method – especially applied in oceanography, but recently also in hydrology – is strongly supported by the experimental mass spectrometric technique, as the amount of ^{3}He produced is extremely small. This is shown here by an example of one litre of water with a ^{3}H activity of 100 TU, which over a period of 20 years has accumulated 5.1×10^{-10} mL STP ($0\,°C$ and 1033 hPa) of ^{3}He (cf. Section II.4).

4.4   NUCLEAR REACTIONS

The overall principle of a *nuclear reaction* is that a nucleus is bombarded by a particle that is in the first instance taken up by the nucleus to form a *compound nucleus*. This does not result in a stable particle. Within an extremely short time the compound nucleus will lose the excess energy by the release of one or more particles, ultimately forming a new nuclide. The bombarding particles may be 'man-made' and originate from a nuclear particle accelerator (electrically charged particles) or a nuclear reactor (high- or low-energy neutrons), or may be from natural origin.

A few examples may serve to show the convention of writing nuclear reactions:

$$^{36}Ar + n \rightarrow p + ^{36}Cl \quad \text{or} \quad ^{36}Ar(n, p)^{36}Cl$$

$$^{40}Ca + n \rightarrow \alpha + ^{37}Ar \quad \text{or} \quad ^{40}Ca(n, \alpha)^{37}Ar$$

$$^{40}Ar + n \rightarrow 2n + ^{39}Ar \quad \text{or} \quad ^{40}Ar(n, 2n)^{39}Ar \tag{4.30}$$

$$^{238}U(n, \gamma)^{239}U, \text{ followed by:}$$

$$^{239}U \rightarrow ^{239}Np(+\beta^-) \rightarrow ^{239}Pu(+\beta^-) \text{ and } ^{239}Pu \text{ (n, fission)}.$$

The last reaction series is the basis of *breeder reactors* (reactors also producing 'fuel').

4.4.1   Natural production

Nature is constantly and intensively performing nuclear reactions. The bombarding particles originate from two sources:

1. From *cosmic radiation*: an avalanche of fragments of atmospheric nuclei, a wide range of particles from photons and electrons to neutrons and mesons. These are primarily produced as secondary and tertiary particles in the upper atmosphere in a process called *spallation*, by very-high-energy cosmic protons and a smaller percentage of helium nuclei from the sun, from further away in our Galaxy and probably from extragalactic

space. Mesons reaching the earth surface have a very high penetrating power and can be observed at great depth below the ground or water surface.

Variations in cosmic radiation occur because the origin is to be found in events such as solar flares and supernova explosions. The sun also indirectly influences the cosmic ray flux in the atmosphere. The fact is that part of the incoming primary particles are deflected away from the earth by the earth's magnetic field, which has not been entirely constant through geologic times. The variable magnetic field of the sun now adds to the variations in the earth's magnetic field, and thus to the shielding capacity of the earth against cosmic particles.

Cosmic radiation produces radioactive nuclei in the atmosphere primarily through two different types of reaction:

(i) high-energy reactions by the secondary particles such as protons, neutrons and He nuclei bombarding atmospheric nuclei (N, O, Ar); these reactions are the origin of the atmospherically abundant radioactive nuclides such as ^{3}H, ^{7}Be, ^{26}Al, ^{32}Si, ^{39}Ar and ^{85}Kr;

(ii) low-energy reactions by thermal neutrons, of which the production of ^{14}C in the air is the best known example.

These reactions are discussed later, together with the specific treatment of the various isotopes (Chapter 6).

2. From *nuclear decay particles*: primarily neutrons from spontaneous decay of nuclides of the uranium, actinium and thorium decay series. These processes occur underground, depending on the U or Th content of the soil minerals (see sections 12.13–12.16 in Volume I of the UNESCO/IAEA series).

4.4.2 Anthropogenic releases of radionuclides

Man has also become able to produce nuclei and even elements at present unknown in nature, on a small scale in the laboratory, but also on a large scale in the natural environment by the test explosion of nuclear weapons.

The anthropogenic releases of radioactivity are related to these nuclear weapons testings as well as to the regular or incidental releases by nuclear (power) industries.

The release of radioactivity related to the testing of nuclear weapons has two main causes, namely:

1. *Fission bombs*, based on the neutron-induced fission reaction of uranium or plutonium:

$$^{235}\text{U} + \text{n} \rightarrow \text{nucleus with } A \approx 85 \text{ to } 100 + \text{nucleus with}$$

$$A \approx 130 \text{ to } 145 + \text{neutrons}$$

and likewise

$$^{239}\text{Pu} + \text{n} \rightarrow \text{nucleus with } A \approx 95 \text{ to } 105 + \text{nucleus with}$$

$$A \approx 132 \text{ to } 142 + \text{neutrons}$$

Well-known examples of radioactive nuclides as products of the asymmetrical mass distribution of the long-living fission products are ^{93}Zr and ^{99}Tc on the one hand, and

^{129}I and ^{137}Cs on the other. Industrial activities relating to nuclear power production and fuel reprocessing may also bring nuclear waste materials into the environment on a small scale, similar to those that are being produced by the bomb explosions.

A special case of nuclear waste release is the occurrence of nuclear reactor incidents. In some cases (Chernobyl in 1986) sufficient radioactivity was injected into the environment to enable detection of certain isotopes in nature, for example, in sediments and ice cores.

2. During nuclear explosions the neutron density in the surrounding air in particular is sufficiently high to **produce radioactive nuclides** that are also being made by nature:

$$^{14}N + n \rightarrow {}^{12}C + {}^{3}H \quad \text{or} \quad {}^{14}N(n, {}^{3}H)^{12}C \tag{4.31}$$

and

$$^{14}N + n \rightarrow {}^{14}C + {}^{1}H \quad \text{or} \quad {}^{14}N(n, p)^{14}C \tag{4.32}$$

^{3}H (tritium) and ^{14}C (radiocarbon) are discussed in detail in Chapter 6.

4.4.3   Radioactive growth

From a mathematical viewpoint the production of radionuclides by nuclear reactions comes to the following. For a constant production rate (P) and simultaneously radioactive decay of the nuclides produced, the net growth rate of the number of nuclides is:

$$\frac{dN}{dt} = P - \lambda N \tag{4.33}$$

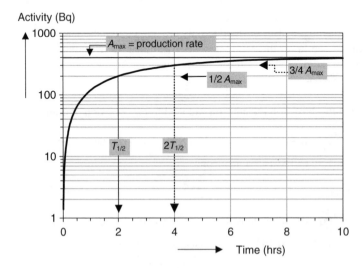

Figure 4.6.   Growth of radioactivity by a constant production rate P of 400 nuclides/sec, resulting in a maximum activity of 400 Bq. The half-life of the produced nuclide is 2 hours.

where N is the number of radioactive nuclei and λ is the decay constant. The net activity increase is thus:

$$\lambda N = A = P(1 - e^{-\lambda t}) \qquad (4.34)$$

As time approaches infinity, a stationary state is being reached in which the radionuclide production and decay are equal. Thus, at $t = \infty$:

$$A_{max} = P \qquad (4.35)$$

This is shown in Fig. 4.6, representing the course of the radioactivity in time. The time required to produce certain fractions of the maximum attainable activity is now:

$A = 1/2\, P = 1/2\, A_{max}$ after one half-life

$A = 3/4\, P$ after two half-lives

$A = 7/8\, P$ after $3T_{1/2}$, and so on.

This means that after three half-lives the maximum attainable activity has effectively been reached.

CHAPTER 5

Natural abundance of the stable isotopes of C, O and H

This chapter is concerned with the natural concentrations of the stable isotopes of carbon, oxygen and hydrogen, with particular attention paid to those compounds relevant in the hydrological cycle. For each isotope separately, we discuss the natural fractionation effects, internationally agreed definitions and standards, and variations in the natural abundances.

In order to help the reader to appreciate isotopic abundance values as they occur in nature, Fig. 5.1 shows some actual isotope ratios and fractionations in a choice of equilibrium systems. Surveys of some practical data for all isotopes concerned are given in the respective sections.

5.1 STABLE CARBON ISOTOPES

5.1.1 The natural abundance

The chemical element carbon has two stable isotopes, ^{12}C and ^{13}C (Table 5.1). Their abundance is about 98.9% and 1.1%, so that the $^{13}C/^{12}C$ ratio is about 0.011. As a result

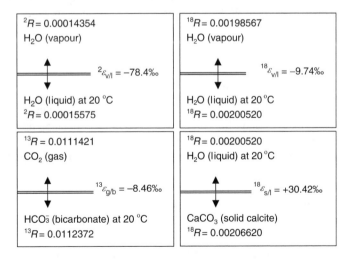

Figure 5.1. Examples of isotope ratios of compounds in isotopic equilibrium and the respective isotope fractionations, as defined in Section 2.3. As an example may serve the calculation of $^{13}\varepsilon_{g/b}$ as $^{13}\alpha_{g/b} - 1$ with $^{13}\alpha_{g/b} = 0.0111421/0.0112372 = 0.99154$.

Table 5.1.   The stable and radioactive isotopes of *carbon*: practical data for the natural abundance, properties, analytical techniques and standards. (Further details are given in Section 5.1 and 6.1 and in the Appendices.)

	^{12}C	^{13}C	^{14}C
Stability	Stable	Stable	Radioactive
Natural abundance	0.989	0.011	$<10^{-12}$
Natural specific activity			<0.25 Bq/gC
Decay mode/daughter			$\beta^-/^{14}N$
Half-life ($T_{1/2}$)			5730 a
Decay constant (λ)			$1.21 \times 10^{-4}a = 1/8267a^{-1}$
Max. β energy			156 keV
Abundance range in hydrological cycle		30‰	0 to 10^{-12}
Reported as		$^{13}\delta$ or $\delta^{13}C$	$^{14}A, ^{14}a, ^{14}\delta$ or $^{14}\Delta$
In		‰	dpm/gC, Bq/gC,% or ‰
Instrument		MS	PGC, LSS, AMS
Analytical medium		CO_2	$CO_2, C_2H_2, CH_4, C_6H_6$, graphite
Usual standard deviation		0.03‰	1‰ to 1% at natural level
International standard		VPDB	Oxalic acid: Ox1, Ox2
With absolute value		0.0112372	13.56 dpm/gC

MS = mass spectrometry, PGC = proportional gas counting, LSS = liquid scintillation spectrometry, AMS = accelerator mass spectrometry.

of several fractionation processes, kinetic as well as equilibrium, the isotope ratio shows a natural variation of almost 100‰.

Fig. 5.2 presents a broad survey of natural abundances of various compounds; at the low-^{13}C end bacterial methane (marsh-gas), and at the high end the bicarbonate fraction of groundwater under special conditions. In the carbonic acid system variations up to 30‰ are normally observed. Wider variations occur in systems in which carbon oxidation or reduction reactions take place, such as the CO_2 (carbonate)–CH_4 (methane) or the $CO_2-(CH_2O)_x$ (carbohydrate) systems.

5.1.2   Carbon isotope fractionations

It is shown later that the presence of *dissolved inorganic carbon* (DIC) in sea-, ground- and surface water presents the possibility for studying gas–water exchange processes and for measuring water transport rates in oceans and in the ground. In connection with studying these phenomena, the stable and radioactive isotopes of carbon and their interactions provide an important contribution, often together with the water chemistry.

In nature, equilibrium carbon isotope effects occur specifically between the phases $CO_2-H_2O-H_2CO_3-CaCO_3$. Values for the fractionations involved depend only on temperature and are obtained from laboratory experiments. A survey is presented in Fig. 5.3 and Table 5.2.

The kinetic fractionation of specific interest is that during carbon dioxide assimilation, that is, the CO_2 uptake by plants. A quantitative estimate shows that the isotope effect, as a result of diffusion of CO_2 through air, cannot explain the fractionation (Section 2.5). The resulting value of $^{13}\alpha$ is 0.9956, so that only $-4.4‰$ of the total assimilation fractionation

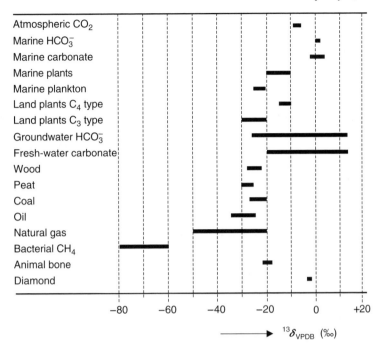

Figure 5.2. General view of $^{13}C/^{12}C$ variations in natural compounds. The ranges are indicative for the materials shown.

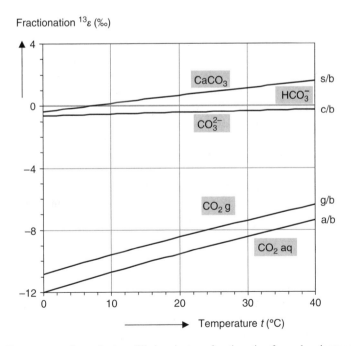

Figure 5.3. Temperature-dependent equilibrium isotope fractionation for carbon isotopes of gaseous CO_2 (g), dissolved CO_2 (a), dissolved carbonate ions (c) and solid carbonate (s) with respect to dissolved HCO_3^- (b). The actual data and equations are given in Table 5.2.

Table 5.2.   Carbon isotope fractionation in the equilibrium system CO_2–HCO_3–CO_3–$CaCO_3$; $^{13}\varepsilon_{y/x}$ represents the fractionation of compound y relative to compound x. (Values for intermediate temperatures may be calculated by linear interpolation (see also Fig. 5.3). $T = t(°C) + 273.15$ K).

t (°C)	$^{13}\varepsilon_{g/b}$ (‰)[1]	$^{13}\varepsilon_{a/g}$ (‰)[2]	$^{13}\varepsilon_{a/b}$ (‰)[3]	$^{13}\varepsilon_{c/b}$ (‰)[4]	$^{13}\varepsilon_{s/b}$ (‰)[5]	$^{13}\varepsilon_{s/g}$ (‰)[6]
0	−10.83	−1.18	−12.00	−0.65	−0.39	+10.55
5	−10.20	−1.16	−11.35	−0.60	−0.11	+10.19
10	−9.60	−1.13	−10.72	−0.54	+0.15	+9.85
15	−9.02	−1.11	−10.12	−0.49	+0.41	+9.52
20	−8.46	−1.09	−9.54	−0.44	+0.66	+9.20
25	−7.92	−1.06	−8.97	−0.39	+0.91	+8.86
30	−7.39	−1.04	−8.42	−0.34	+1.14	+8.60
35	−6.88	−1.02	−7.90	−0.29	+1.37	+8.31
40	−6.39	−1.00	−7.39	−0.25	+1.59	+8.03

According to References and selected papers given at the end of this volume:

[1] $^{13}\varepsilon_{g/b} = -9483/T + 23.89‰$ (5.1)

[2] $^{13}\varepsilon_{a/g} = -373/T + 0.19‰$ (5.2)

[3] $^{13}\varepsilon_{a/b} = -9866/T + 24.12‰$ (5.3)

[4] $^{13}\varepsilon_{c/b} = -867/T + 2.52‰$ (5.4)

[5] $^{13}\varepsilon_{s/g} = -4232/T + 15.10‰$ (5.5)

[6] $^{13}\varepsilon_{s/g} = +5380/T - 9.15‰$ (5.6)

g = gaseous CO_2, a = dissolved CO_2, b = dissolved HCO_3^-, c = dissolved CO_3^{2-} ions, s = solid calcite.

in favour of ^{12}C can be explained by the diffusion. The remaining −13.6‰, therefore, has to be found at the surface of the liquid phase and in the subsequent biochemical process.

Another kinetic process occurring in the soil is the bacterial decomposition of organic matter to form methane (CH_4). Here the largest fractionation amounts to about −55‰. In this process CO_2 is simultaneously produced with a fractionation of +25‰ relative to $^{13}\delta \approx -25‰$ for organic matter, consequently resulting in a $^{13}\delta$ value for this CO_2 of about 0‰.

A special problem is the fractionation during uptake and release of CO_2 by seawater. This fractionation is included in calculations relating to global ^{13}C modelling.

Strictly speaking, the difference in isotopic composition between gaseous CO_2 and the dissolved inorganic carbon content (DIC) of water cannot be addressed by *isotope fractionation between CO_2 and DIC*.

Isotope fractionation is the phenomenon that, due to a physical or chemical isotope exchange process, a difference occurs between the isotopic composition *of two compounds*, while seawater carbon consists of three fractions: that is, dissolved CO_2 (H_2CO_3 is hardly present), dissolved HCO_3^-, and dissolved CO_3^{2-}. All these fractions are fractionated isotopically with respect to each other.

The ^{13}R value of DIC is:

$$^{13}R_{DIC} = \frac{[CO_2 aq]^{13}R_{CO_{2aq}} + [HCO_3^-]^{13}R_{HCO_3} + [CO_3^{2-}]^{13}R_{CO_3}}{[CO_2 aq] + [HCO_3^-] + [CO_3^{2-}]}$$

or in terms of the respective fractionations:

$$^{13}R_{DIC} = \frac{a\,^{13}\alpha_{a/b} + b + c\,^{13}\alpha_{c/b}}{C_T}\,^{13}R_b \tag{5.7}$$

or in terms of δ values:

$$^{13}\delta_{DIC} = \frac{a\,^{13}\alpha_{a/b} + b + c\,^{13}\alpha_{c/b}}{C_T}\,^{13}\delta_b + \frac{a\,^{13}\varepsilon_{a/b} + c\,^{13}\varepsilon_{c/b}}{C_T} \tag{5.8}$$

where the brackets indicate the respective concentrations, which are also denoted by a (acid), b (bicarbonate) and c (carbonate ions), so that $a + b + c = C_T$. The α values are given in Table 5.2 ($\alpha = \varepsilon + 1$). The chemical fractions are quantitatively treated in Appendix III.

In oceanic studies values for kinetic fractionations and DIC play an important role, especially that for CO_2 uptake by seawater:

$$^{13}\varepsilon_k(\text{air} \Rightarrow \text{sea}) = {}^{13}\varepsilon_k(\text{atm } CO_2 \text{ to } CO_2 \text{ taken up}) = -2.0 \pm 0.2\%o \tag{5.9}$$

(We remind the reader that these $^{13}\varepsilon_k$ values, as well as those given later, are for *kinetic* fractionations and *do not* refer to Table 5.2).

The kinetic fractionation during CO_2 release by the ocean is discussed in detail in Volume I of the UNESCO/IAEA series. The fractionation of released CO_2 with respect to DIC is:

$$^{13}\varepsilon_k(\text{sea} \Rightarrow \text{air}) = {}^{13}\varepsilon_k(\text{DIC rel. to } CO_2 \text{ released}) = -10.3 \pm 0.3\%o \tag{5.10}$$

5.1.3   Reporting ^{13}C variations and the ^{13}C standard

As described in Section 2.8, isotopic compositions expressed as $^{13}\delta$ values are related to those of specific reference materials. The history of these isotopic reference materials is complicated; an account is given in chapter 7 of Volume I of the UNESCO/IAEA series.

By international agreement, PDB was used as the *primary carbon standard* material. PDB (Pee Dee Belemnite) was the carbonate from a certain (marine) belemnite found in the Cretaceous Pee Dee formation of North America. This material has long been exhausted. The US National Bureau of Standards therefore distributed another (marine) limestone, NBS19, of which $^{13}\delta$ has been accurately established in relation to PDB. In this way the PDB scale is indirectly established. Based on this comparison, in 1983 an IAEA panel adopted the standard to define the new *VPDB (Vienna PDB) scale* as:

$$^{13}\delta_{NBS19/VPDB} = +1.95\%o \tag{5.11}$$

Henceforth all $^{13}\delta$ values are reported relative to VPDB, unless stated otherwise.

More details on measurement and calculation procedures, and on reference samples, are given in Appendix II.

5.1.4   Survey of natural ^{13}C variations

Certain aspects of natural $^{13}\delta$ variations will be discussed in more detail in the other chapters. Here we restrict ourselves to a general survey, particularly with regard to the hydrological cycle (Fig. 5.4).

5.1.4.1   *Atmospheric CO_2*

The least depleted atmospheric CO_2 originally had $^{13}\delta$ values near $-7\%o$; since the nineteenth century this value has undergone relatively large changes. In general, high values are observed in oceanic air far removed from continental influences and occur in combination with minimal CO_2 concentrations. More negative $^{13}\delta$ values are found in continental air and are due to an admixture of CO_2 of biospheric and anthropogenic origin ($^{13}\delta \approx -25\%o$), in part from the decay of plant material, in part from the combustion of fossil fuels.

5.1.4.2   *Seawater and marine carbonate*

Atmospheric CO_2 appears to be nearly in isotopic equilibrium with the oceanic dissolved bicarbonate. The $^{13}\delta(HCO_3^-)$ values in the ocean are about $+1$ to $+1.5\%o$, in agreement with the equilibrium fractionation $\varepsilon_{g/b}$ at temperatures between 15 and 20 °C (Table 5.2). According to the equilibrium fractionation $\varepsilon_{s/b}$, we should expect calcite slowly precipitating in equilibrium with oceanic bicarbonate to have $^{13}\delta$ values of $+2.0$ to $+2.5\%o$. This is indeed the normal range found for recent marine carbonates, and for marine and brackish-water shells.

5.1.4.3   *Vegetation and soil CO_2*

Plant carbon has a lower ^{13}C content than the atmospheric CO_2 from which it was formed. The fractionation which occurs during CO_2 uptake and photosynthesis depends on the type of plant, and on the climatic and ecological conditions.

The dominant modes of photosynthesis give rise to strongly differing degrees of fractionation. The Hatch-Slack photosynthetic pathway (C_4) results in $^{13}\delta$ figures of -10 to $-15\%o$, and is primarily represented by certain grains and desert grasses (sugar cane, corn). In temperate climates most plants employ the Calvin mechanism (C_3), producing $^{13}\delta$ values in the range $-26 \pm 3\%o$. A third type of metabolism, the Crassulacean Acid Metabolism (CAM), produces a large spread of $^{13}\delta$ values around $-17\%o$.

The CO_2 content of the soil atmosphere can be orders of magnitude larger than that for the free atmosphere. The additional CO_2 is formed in the soil by decay of plant remains and by root respiration and, consequently, has $^{13}\delta$ values which centre around $-25\%o$ in temperate climates where Calvin plants dominate.

5.1.4.4   *Fossil fuel*

As complicated biogeochemical processes are involved in the degradation of terrestrial and marine plant material ultimately into coal, oil and natural gas, the range of $^{13}\delta$ values for these fossil fuels is larger, extending to more negative values especially in biogenic methane (Fig. 5.4). The global average of CO_2 from the combustion of these fuels is estimated to be about $-27\%o$.

5.1.4.5   *Global carbon cycle*

Biospheric carbon has a direct influence on $^{13}\delta$ of atmospheric CO_2. The large uptake of CO_2 by the global biosphere during summer, and the equal release of CO_2 during winter,

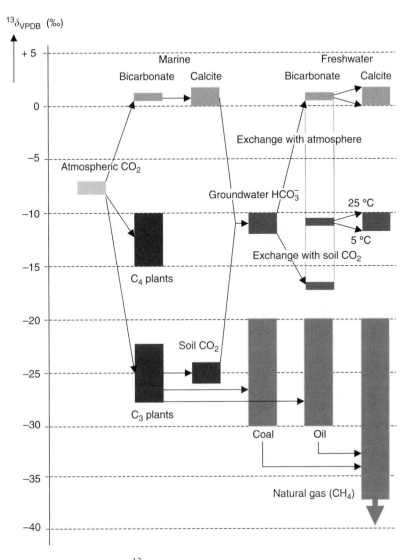

Figure 5.4. Schematic survey of $^{13}\delta$ variations in nature, especially of compounds relevant in the hydrological cycle.

causes a seasonal variation in the atmospheric CO_2 concentration as well as in $^{13}\delta$. The simple mixing of CO_2 from these two components, atmospheric CO_2 (atm) and biospheric CO_2 (bio), is represented by the equation (cf. Eq. 3.2):

$$(^{13}\delta_{atm} + \Delta^{13}\delta)(C_{atm} + \Delta C_{bio}) = {}^{13}\delta_{atm}C_{atm} + {}^{13}\delta_{bio}\Delta C_{bio} \qquad (5.12)$$

where C is the CO_2 concentration, ΔC the biospheric addition, and $\Delta^{13}\delta$ the variation in the δ value.

Numerically this reduces to a periodic (seasonal) variation of

$$\frac{\Delta^{13}\delta}{\Delta C_{bio}} = \frac{{}^{13}\delta_{bio} - {}^{13}\delta_{atm}}{C_{atm} + \Delta C_{bio}} = \frac{-25 + 8.00}{370} \approx -0.05 \text{ ‰ per ppm of } CO_2 \quad (5.13)$$

Superimposed on this phenomenon is a gradual increase in CO_2 concentration and an accompanying decrease in $^{13}\delta$ from the emission of fossil fuel CO_2. The trends are shown in Fig. 5.5 and can be approximated by:

$$\Delta^{13}\delta / \Delta CO_2 = -0.015\text{‰/ppm} \quad \text{or} \quad \Delta^{13}\delta = -0.025\text{‰/year} \quad (5.14)$$

at a CO_2 concentration of 370 ppm and $^{13}\delta = -8.00\text{‰}$ over the Northern Hemisphere, valid for January 2000.

The smaller ‰/ppm value of Eq. 5.14 compared to Eq. 5.13 shows that the long-term trend is not due to simple addition and mixing of additional CO_2 in the atmosphere. The large oceanic DIC reservoir levels out the purely atmospheric mixing effect through isotope exchange.

5.1.4.6 *Groundwater and riverwater*
Soil CO_2 is important for establishing the dissolved inorganic carbon content of ground-water. After dissolution of this CO_2 the infiltrating rain water is able to dissolve the soil

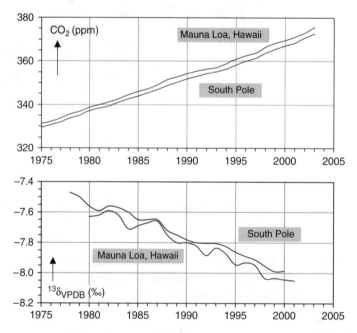

Figure 5.5.   Trends of the concentration and $^{13}\delta$ of atmospheric CO_2 for air samples collected on top of Mauna Loa volcano, Hawaii, and at the South Pole. The seasonal variations have been removed from the original data. The dates refer to the mean values for each year (concentration data from Keeling and Whorf, ^{13}C data from CIO (Groningen) and the NOAA website (J. White), see References and selected papers).

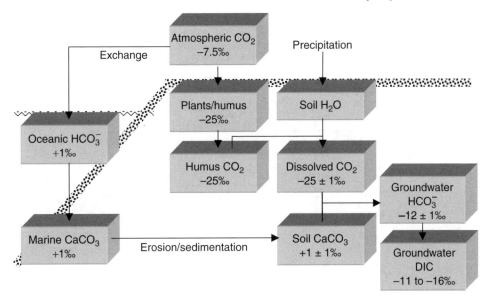

Figure 5.6.   Schematic representation of the formation of dissolved inorganic carbon in groundwater from soil carbonate and soil CO_2. This is the main process responsible for the carbonate content of groundwater and the consecutive components of the water cycle. Generally, dissolved bicarbonate is by far the largest component. The ‰ values referring to the respective $^{13}\delta$ have been kept simple for the sake of clarity. DIC is the dissolved inorganic carbon content of the water; that is, HCO_3^-, $CO_2(aq)$ and CO_3^{2-}.

limestone (Fig. 5.6):

$$CO_2 + H_2O + CaCO_3 \rightarrow Ca^{2+} + 2HCO_3^- \qquad (5.15)$$

Because limestone is generally of marine origin ($^{13}\delta \approx +1‰$), this process results in $^{13}\delta$ for the dissolved bicarbonate of about -11 to $-12‰$ (in temperate climates).

In the soil the HCO_3^- first formed exchanges with the often present excess of gaseous CO_2, *ultimately* resulting in $^{13}\delta(HCO_3^-) = {}^{13}\delta(\text{soil } CO_2) + {}^{13}\varepsilon_{b/g} \approx -25‰ + 9‰ = -16‰$ (Fig. 5.4). Consequently, $^{13}\delta(HCO_3^-)$ values significantly outside the range of -11 to $-12‰$ are observed in soil water as well as in fresh surface water such as rivers and lakes. In surface waters such as lakes, ^{13}C enrichment of dissolved inorganic carbon can be caused by isotope exchange with atmospheric CO_2 ($^{13}\delta \approx -7.5‰$), *ultimately* resulting in values of $^{13}\delta + {}^{13}\varepsilon_{b/g} = -7.5‰ + 9‰ = +1.5‰$, identical to oceanic values. Consequently, freshwater carbonate minerals may have ‘marine’ $^{13}\delta$ values. In these cases the marine character of the carbonate is determined by $^{18}\delta$ (Section 5.3).

In addition to HCO_3^-, natural waters contain variable concentrations of CO_2 with the effect that the $^{13}\delta$ value of DIC is lower (more negative) than that of the bicarbonate fraction alone; in groundwater, and in stream and river waters derived from groundwater, the $^{13}\delta$ (DIC) values are generally in the range of -12 to $-15‰$ (Fig. 5.7).

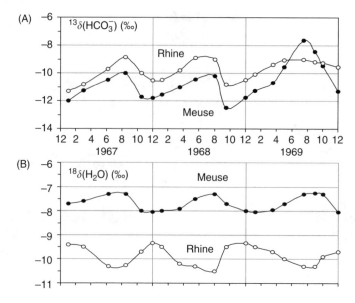

Figure 5.7.   A three-year observation of the isotopic composition of water from the North Western European rivers Rhine and Meuse in the Netherlands: (A) $^{13}\delta$ values for the dissolved bicarbonate fraction, showing normal groundwater-values during winter and relatively high values in summer, probably because of isotopic exchange of the surface water bicarbonate with atmospheric CO_2. (B) $^{18}\delta$ values, where the River Meuse is showing the average value and the seasonal variations of $^{18}\delta$ in the precipitation; high values in summer, low during winter. During early spring and summer the Rhine receives melt water from the Swiss Alps with relatively low $^{18}\delta$ because of the high-altitude precipitation.

5.2   STABLE OXYGEN ISOTOPES

5.2.1   The natural abundance

The chemical element oxygen has three stable isotopes, ^{16}O, ^{17}O and ^{18}O, with natural abundances of 99.76, 0.035 and 0.2%, respectively. ^{17}O concentrations provide little information on the hydrological cycle in the strict sense above that which can be gained from the more abundant and, consequently, more accurately measurable ^{18}O variations (Section 2.7). We shall, therefore, focus our attention here on the $^{18}O/^{16}O$ ratio (≈ 0.0020).

Table 5.3 contains some analytical and technical data concerning the abundance and measurement of $^{18}O/^{16}O$ ratios.

Values of $^{18}\delta$ show natural variations within a range of almost 100‰ (Fig. 5.8). ^{18}O is often enriched in (saline) lakes subjected to a high degree of evaporation, while high-altitude and cold-climate precipitation, especially in the Antarctic, is low in ^{18}O. Generally, in the hydrological cycle in temperate climates we are confronted with a range of $^{18}\delta$ not exceeding 30‰.

5.2.2   Oxygen isotope fractionations

The isotope effects to be discussed are within the system H_2O (vapour)–H_2O (liquid)–$CaCO_3$. The equilibrium fractionation values have been determined by laboratory

Table 5.3.   The stable isotopes of *oxygen*: practical data for the natural abundance, properties, analytical techniques and standards. (Further details are given in Section 5.2, and in Appendices I and II.)

	^{16}O	^{17}O	^{18}O
Stability	Stable	Stable	Stable
Natural abundance	0.9976	0.00038	0.00205
Abundance range in hydrological cycle		15‰	30‰
Reported as		$^{17}\delta$ or $\delta^{17}O$	$^{18}\delta$ or $\delta^{18}O$
In		‰	‰
Instrument		MS	MS
Analytical medium		O_2	CO_2 or O_2
Usual standard deviation			0.05‰
International standard		.	VSMOW for water, VPDB for carbonate, etc.
With absolute value			VSMOW: 0.0020052 VPDB: 0.0020672

MS = mass spectrometry.

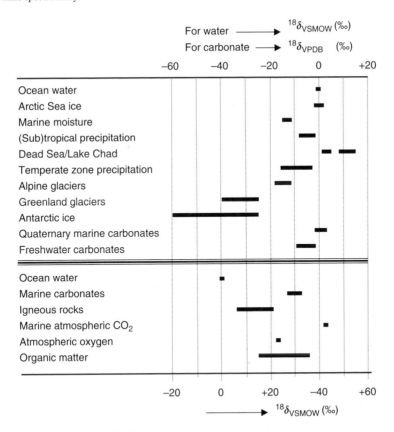

Figure 5.8.   General view of $^{18}O/^{16}O$ variations in natural compounds. The ranges are indicative for the majority of materials shown. The relation between the VPDB and VSMOW scales is given in Section 5.2.3 and Fig. 5.10.

experiments. Fig. 5.1 shows some actual isotope ratios, and Fig. 5.9 and Table 5.4 present a summary of the temperature-dependent equilibrium isotope effects.

Equilibrium fractionations determined in the laboratory are also found in nature. The most striking observation is that the carbonate shells of many molluscs appear to have been formed in isotopic equilibrium with seawater.

The *palaeotemperature scale* is presented by Eq. 5.20. This relation is deduced from ^{18}O measurements of carbonate laid down by marine shell animals at known temperatures and water isotopic compositions.

Kinetic effects are observed during the evaporation of ocean water, as oceanic vapour is isotopically lighter than would result from equilibrium fractionation alone. The natural isotope effect for oxygen ($\approx -12\%_o$) is smaller than could result from fractionation by diffusion; laboratory measurements resulted in $^{18}\varepsilon_d = -27.2 \pm 0.5\%_o$. This experimental value is smaller than would be expected from Eq. 2.25 ($-31.3\%_o$), which may be explained by the water molecules forming clusters of larger mass in the vapour phase. Furthermore, evaporation of ocean water includes sea spray, from which water droplets evaporate completely without fractionation, thus reducing the natural isotope effect.

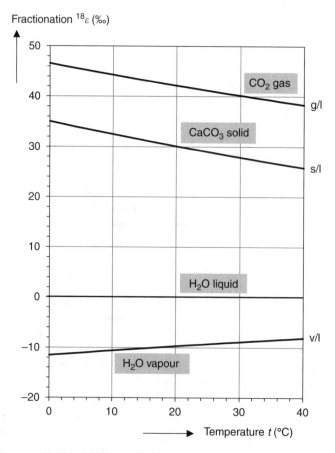

Figure 5.9.   Temperature-dependent equilibrium fractionations for oxygen isotopes of water vapour (v), gaseous CO_2 (g) and solid calcite (s) with respect to liquid water (l).

Table 5.4. Oxygen isotope fractionation in the equilibrium system CO_2–H_2O–$CaCO_3$; $^{18}\varepsilon_{y/x}$ represents the fractionation of compound y relative to compound x, and is approximately equal to $\delta_y - \delta_x$. (Values for intermediate temperatures may be calculated by linear interpolation (Fig. 5.9); $T = t\,(°C) + 273.15\ \text{K}$.)

t (°C)	$^{18}\varepsilon_{v/l}$ (‰)[1]	$^{18}\varepsilon_{s/l}$ (‰)[2]	$^{18}\varepsilon_{g/l}$ (‰)[3]	$^{18}\varepsilon_{g/lg}$ (‰)[4]	$^{18}\varepsilon_{sg/lg}$ (‰)[5]
0	−11.55	+34.68	+46.56	+5.19	+3.98
5	−11.07	+33.39	+45.40	+4.08	+2.72
10	−10.60	+32.14	+44.28	+3.01	+1.51
15	−10.15	+30.94	+43.20	+1.97	+0.34
20	−9.71	+29.77	+42.16	+0.97	−0.79
25	−9.29	+28.65	+41.15	0	−1.88
30	−8.89	+27.56	+40.18	−0.93	−2.93
35	−8.49	+26.51	+39.24	−1.84	−3.96
40	−8.11	+25.49	+38.33	−2.71	−4.94

According to References and selected papers given at the end of this volume:

[1] $\ln\ {}^{18}\alpha_{v/l} = -1137/T^2 + 0.4156/T + 0.0020667$ (5.16a)

     Exponential : $^{18}\alpha_{v/l} = 1.0157\exp(-7.430/T)$ (5.16b)

         $1/T :\ {}^{18}\varepsilon_{v/l} = -7356/T + 15.38‰$ (5.16c)

[2] $^{18}\varepsilon_{s/l} = 19\,668/T - 37.32‰$ (5.17)

[3] $^{18}\varepsilon_{g/l} = 17\,604/T - 17.89‰$ (5.18)

[4] $^{18}\varepsilon_{g/lg} = 16\,909/T - 56.71‰$ (5.19)

     independent of the salt content of the solution.

[5]   Historically famous palaeotemperature scale:

$$t(°C) = 16.5 - 4.3(^{18}\delta_s - {}^{18}\delta_w) + 0.14(^{18}\delta_s - {}^{18}\delta_w)^2 \qquad (5.20)$$

$^{18}\delta$ refers to CO_2 prepared from solid carbonate with 95% H_3PO_4 at 25 °C, and $^{18}\delta_w$ to CO_2 in isotopic equilibrium with water at 25 °C, both relative to VPDB-CO_2; sg is obtained from s by applying the fractionation for the CO_2 production at 25 °C:

$^{18}\varepsilon_{sg/s} = +10.25‰$ (5.21)

$^{18}\varepsilon_{sg/lg} = 19\,082/T - 65.88‰$ (5.22)

According to reference [1]: $^{18}\varepsilon_{i/l} = +3.5‰\ (0\ °C)$ (5.23)

$^{18}\varepsilon_{i/v} = +15.2‰\ (0\ °C);\quad {}^{18}\varepsilon_{i/v} = +16.6‰\ (-10\ °C)$ (5.24)

l = liquid H_2O, v = H_2O vapour, i = ice , g = gaseous CO_2, s = solid $CaCO_3$, lg = CO_2 (g) in isotopic equilibrium with H_2O (l) at 25 °C, sg = CO_2 (g) from $CaCO_3$ (s) by 95% H_3PO_4 at 25 °C.

5.2.3   Reporting ^{18}O variations and the ^{18}O standards

Section 7.2.3 in Volume I of the UNESCO/IAEA series gives a complete historic account of the complicated issue of ^{18}O standards. Serving as *a water standard*, the International Atomic Energy Agency (IAEA), Section of Isotope Hydrology, in Vienna, Austria and the

US National Institute of Standards and Technology (NIST, the former NBS) have available for distribution batches of carefully prepared and well preserved standard mean ocean water for use as a standard for ^{18}O as well as for ^2H.

At present two standard materials are available for reporting $^{18}\delta$ values; one for water samples and one for carbonates. This situation arises from the practical fact that neither the isotope measurements on water nor those on carbonates are performed on the original material itself, but are made on gaseous CO_2 which has reacted with or is derived from the sample.

The laboratory analysis of $^{18}O/^{16}O$ in water is performed by equilibrating a water sample with CO_2 of known isotopic composition at 25 °C (Appendix I), followed by mass spectrometric analysis of this equilibrated CO_2 (Appendix II). This equilibration is generally carried out on batches of water samples, consisting of the unknown samples (x) and the standard or one or more reference samples. After the correction discussed in Appendix II is made, it is irrelevant whether the water samples themselves are being reported or the CO_2 samples obtained after equilibration with the water, provided sample and standard are treated similarly:

$$^{18}\delta_{x/VSMOW} = {}^{18}\delta_{xg/VSMOWg} \tag{5.25}$$

where g refers to the equilibrated and analysed CO_2.

For reporting carbonate $^{18}\delta$ values, a carbonate reference material, NBS19, has been introduced, which defines the VPDB scale by:

$$^{18}\delta_{NBS19/VPDB} = -2.20\text{‰} \tag{5.26}$$

The carbonate itself is not analysed for $^{18}\delta$, but rather the CO_2 prepared according to a standard procedure which involves treatment in vacuum with 95% (or 100%) orthophosphoric acid at 25 °C. If samples and reference are treated and corrected similarly,

$$^{18}\delta_{x/VPDB} = {}^{18}\delta_{xg/VPDBg} \tag{5.27}$$

where g refers to the prepared and analysed CO_2, so that neither the fractionation between the carbonate and the CO_2 prepared from it (Table 5.4), nor the reaction temperature need to be known.

$^{18}\delta$ *values for water samples* are referred to VSMOW (i.e. Vienna Standard Mean Ocean Water), replacing the original SMOW in fixing the zero point for the $^{18}\delta$ scale.

$^{18}\delta$ *values for carbonates* are given with reference to the same VPDB (i.e. Vienna PeeDee Belemnite) calcite used for ^{13}C (Section 5.1.3).

$^{18}\delta$ *values of (atmospheric) CO_2* are generally reported relative to PDB-CO_2, rather than to PDB itself.

The relations between VSMOW and VPDB (and a few secondary standards, not relevant in this discussion, see Volume I) are visualised in Fig. 5.10. These are of interest in isotope studies on silicates, oxides, carbonates, CO_2 and organic matter, and their relationship with water.

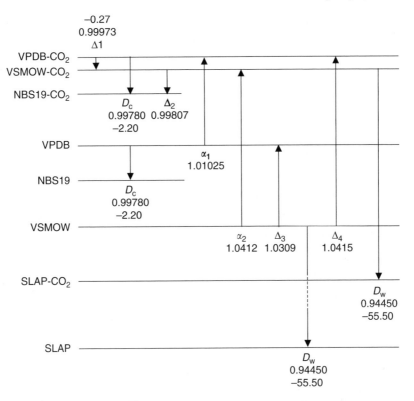

Figure 5.10. Relations between ^{18}O reference and intercomparison samples with respect to VPDB and VSMOW. VPDB-CO_2 refers to CO_2 prepared from hypothetical VPDB by treatment with H_3PO_4 (95%) at 25 °C, and VSMOW-CO_2 to CO_2 equilibrated with VSMOW at 25 °C. The vertical scale is indicative and not entirely proportional to real numbers.

Δ_1: difference between VPDB-CO_2 and SMOW-CO_2 (−0.22‰), plus the difference between SMOW(CO_2) and VSMOW(CO_2)(−0.05‰)
D_c: defined value of NBS19 relative to VPDB (Eq. 5.26)
Δ_2: = $D_c − \Delta_1$
α_1: fractionation during the production of CO_2 from $CaCO_3$ by treatment with 100% H_3PO_4 at 25 °C
α_2: average of 3 independent methods applied by 4 different laboratories
Δ_3: from α_2, α_1 and Δ_1; the exact figure is 1.03086, in agreement with $\alpha_2 = 1.04115$
Δ_4: from Δ_3 and α_1

The conversion equations are in general:

$$^{18}\delta_{lower} = \alpha_i \, {}^{18}\delta_{upper} + \varepsilon_i \tag{5.28}$$

where $\alpha_i = \Delta$, D or α, and 'lower' and 'upper' refer to the levels of the reference samples in the scheme. For example:

$$^{18}\delta(\text{x rel. to NBS19}) = {}^{18}\alpha(\text{VPDB rel. to NBS19}) \times {}^{18}\delta(\text{x rel. to VPDB})$$

$$+ ({}^{18}\alpha - 1)$$

$$= (1/0.99780) \times {}^{18}\delta(\text{x rel. to VPDB}) + 2.20‰$$

The relations between the VPDB, VPDB-CO$_2$ (VPBDg), VSMOW and VSMOW-CO$_2$ (VSMOWg) scales are derived from the equations given in Table 5.4, according to Eq. 5.28:

$$^{18}\delta_{x/VSMOW} = 1.03086 \,^{18}\delta_{x/VPDB} + 30.86‰ \tag{5.29}$$

$$^{18}\delta_{x/VSMOW} = 1.04143 \,^{18}\delta_{x/VPDBg} + 41.43‰ \tag{5.30}$$

$$^{18}\delta_{x/VSMOWg} = 1.00027 \,^{18}\delta_{x/VPDBg} + 0.27‰ \tag{5.31}$$

More details on the measurement and calculation procedures are given in Appendix II.

5.2.4   Survey of natural ^{18}O variations

$^{18}\delta$ variations throughout the hydrologic cycle are discussed in detail in Chapters 7–9. Here we present only a broad survey (Fig. 5.11).

5.2.4.1   *Seawater*
The oceans form the largest global water reservoir. The ^{18}O content in the surface layer is rather uniform, varying between +0.5 and −0.5‰. Only in tropical and polar regions are larger deviations observed. In tropical regions more positive values are caused by strong evaporation: $^{18}\delta$ of water in the Mediterranean amounts to +2‰ ($^2\delta$ = +10‰). In polar regions more negative values originate from water melting from isotopically light snow and ice.

If ocean water were evaporating under equilibrium conditions, the resulting vapour would be 8 to 10‰ depleted in ^{18}O, depending on temperature. However, oceanic vapour appears to have an average $^{18}\delta$ value of −12 to −13‰, which must partly be due to kinetic fractionation. The relative humidity of the air and the evaporation temperature influence the amount of non-equilibrium fractionation occurring (UNESCO/IAEA Volume II on Atmospheric Water).

5.2.4.2   *Precipitation*
The transformation of atmospheric water vapour to precipitation depends on so many climatological and local factors that the $^{18}\delta$ variations in precipitation over the globe are very large. As a general rule, in equatorial regions $^{18}\delta$ becomes more negative the further the rain is removed from the main source of the vapour. In the Arctic and the Antarctic, $^{18}\delta$ of the ice can be as low as −50‰. This gradual ^{18}O depletion model is schematically shown in Fig. 5.12.

The various effects causing regional and temporal variations in $^{18}\delta$ of precipitation are discussed in detail in Chapter 7.

With regard to the various quantitative effects we can distinguish, in summary, between:

1. *Latitudinal effect*, with lower $^{18}\delta$ values at increasing latitude.
2. *Continental effect*, with increasingly negative $^{18}\delta$ values for precipitation the more inland.
3. *Altitude effect*, with decreasing $^{18}\delta$ in precipitation at higher altitude.
4. *Seasonal effect* (in regions with temperate climate), with more negative $^{18}\delta$ values during winter.
5. *Amount effect*, with more negative $^{18}\delta$ values in rain during heavy storms.

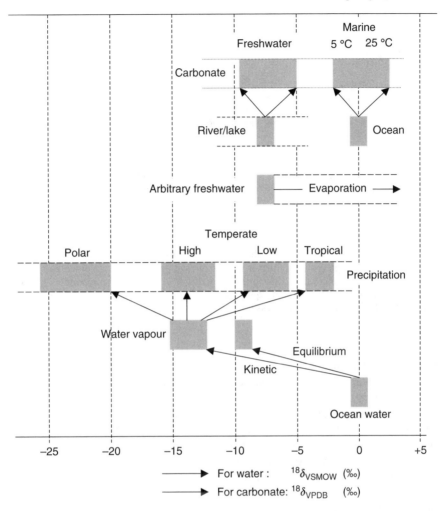

Figure 5.11.   Schematic survey of natural $^{18}\delta$ variations in nature, especially those relevant in the hydrological cycle. The marine vapour gradually becomes depleted in ^{18}O during transport to higher latitudes (Fig. 5.12). Evaporation of surface water may cause the water to become enriched in ^{18}O. Finally, the formation of solid carbonate results in a shift in $^{18}\delta$ depending on temperature (cf. Fig. 5.9).

5.2.4.3   *Surface water*

Data on $^{18}\delta$ variations in river water are given in Chapter 8. The seasonal variation, with relatively high values in summer, is characteristic for precipitation in temperate regions. The basis of the curves in Fig. 5.7 are the average $^{18}\delta$ values for precipitation and groundwater in the recharge areas, namely: North Western Europe (Meuse) and Switzerland/Southern Germany (Rhine), the latter with a large transport of relatively isotopically light meltwater in spring.

Evaporation, especially in tropical and semi-arid regions, causes ^{18}O enrichment in surface waters. This results, for example, in $^{18}\delta$ for the river Nile to be +3 to +4‰, and for certain lakes to be as high as +20‰ (Chapter 7).

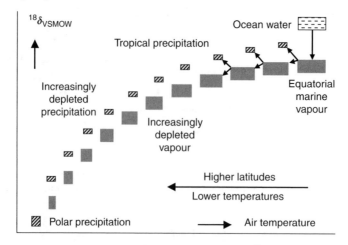

Figure 5.12. Schematic representation of the gradual depletion in ^{18}O for atmospheric water vapour and the condensing precipitation, the further the vapour and the precipitation process are removed from the main source of the vapour: the tropical marine belt. At lower temperatures the isotope fractionation between water vapour and liquid water is larger, counteracting, but only diminishing, the effect of the Rayleigh depletion process (cf. Fig. 3.4 and Section 3.2.1).

5.2.5 Climatic variations

As was pointed out earlier, the slow precipitation of calcium carbonate is a process during which carbonate and water are in isotopic equilibrium. The ^{18}O content of the carbonate is, therefore, primarily determined by that of the water. The second determining factor is the temperature, as indicated in Fig. 5.9. Consequently, we can in principle deduce the water temperature from $^{18}\delta$ of carbonate samples in marine sediment cores, provided $^{18}\delta$ for the water is known ($^{18}\delta$ for seawater $\approx 0‰$). This was originally believed to be the basis of the ^{18}O palaeothermometry of fossil marine shells.

Present-day opinions assume a varying $^{18}\delta$ for the surface ocean water during glacial/interglacial transitions, due to varying amounts of accumulated ice with low $^{18}\delta$ as polar ice caps (Fig. 5.13).

As a realistic order of magnitude, an estimated amount of 5.10^5 km^3 of ice ($= V_{ice}$) laid down especially on the northern polar ice cap during the last ice age, with an average $^{18}\delta$ value of $-20‰$, changes the $^{18}\delta$ value (at present $= 0‰$) of the 10^7 km^3 of ocean water ($= V_{ocean}$) by $+1‰$. This as simply deduced from the mass balance (Eq. 3.2):

$$V_{present\ ocean} \times {}^{18}\delta_{present\ seawater} = V_{ice\text{-}age\ ocean} \times {}^{18}\delta_{ice\text{-}age\ seawater}$$

$$+ V_{ice\ caps} \times {}^{18}\delta_{ice} \tag{5.32}$$

Another spectacular application of isotope variations in nature is the deduction of past climatic changes from $^{18}O/^{16}O$ or $^2H/^1H$ ratios in polar ice cores. If the process of gradual precipitation isotope depletion is studied in detail as a function of latitude, and thus of air temperature, a relation can be derived (Section 7.1.2.1) which leads to the temperature dependence:

$$d^{18}\delta/dt \approx +0.7‰/°C \tag{5.33}$$

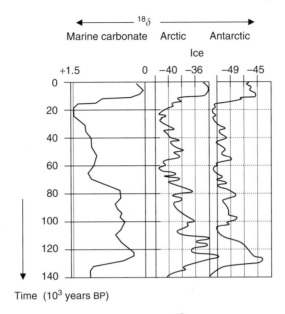

Figure 5.13.   Palaeotemperature records represented by $^{18}\delta$ variations in time in the $CaCO_3$ fraction from foraminifera shells in deep-sea core sediments, in glacier ice of a Northern Greenland glacier (at Summit), and in glacier ice of an Antarctic ice core (at Vostok), respectively from left to right. In each record the left-hand side indicates the lower temperatures; at 20 000 years BP each curve shows the most recent glacial maximum. The upper 10 000 years contain the present warm interglacial, the Holocene.

By this relation it is possible to translate isotope variations into climatic variations during geologic times. Records have been obtained from ice cores in Greenland and the Antarctic, which show the alternation of low-$^{18}\delta$ (or low-$^{2}\delta$) and high-$^{18}\delta$ (or high-$^{2}\delta$) periods, or respectively cold and warm periods (Eq. 5.33).

5.3   RELATION BETWEEN ^{13}C AND ^{18}O VARIATIONS IN H_2O, HCO_3^- AND CO_3^{2-}

Differences and the relations between the various natural water–carbonate systems can be conveniently displayed by considering both $^{18}\delta$ of the water and $^{13}\delta$ of the dissolved bicarbonate. Fig. 5.14 is a schematic representation of three waters of realistic isotopic composition, each provided with the temperature-dependent range of calcite precipitated under equilibrium conditions. This figure essentially combines the graphs from Figs 5.4 and 5.11.

From an isotopic point of view, the four common types of water are:

1. *Seawater*, with $^{18}\delta$ values around 0‰ (by definition) at present; the carbonate range is that of recent marine carbonates. Because of changing glacial and interglacial periods the $^{18}\delta$ of ocean water varied in the past. Also, the $^{18}\delta$ values for marine limestone have increased during the course of geological time, while the $^{13}\delta$ values have essentially remained the same.

2. *Ground- and river water*, with an arbitrarily chosen value of $^{18}\delta$. In fresh-water bicarbonate $^{13}\delta$ generally is around -11 to $-12\%_o$. The isotopic compositions of fresh-water carbonates derived from this water may result from the known equilibrium fractionations (Tables 5.2 and 5.4), in a similar way as is indicated for marine carbonate.

3. *Stagnant surface* or *lake water* can be subjected to processes which alter the isotopic composition. Provided the water has a sufficient residence time in the basin, isotope exchange will cause the ^{13}C content to reach isotopic equilibrium with atmospheric CO_2; then $^{13}\delta$ equals that in the ocean. Carbonate $^{13}\delta$ and $^{18}\delta$ are related to, respectively, HCO_3^- and H_2O as indicated for the marine values. The ^{18}O content of the water, especially in warmer climates, will change towards less negative or even positive values due to evaporation.

4. *Estuarine water* has intermediate values of $^{13}\delta(HCO_3^-)$ or $^{13}\delta(DIC)$ and $^{18}\delta(H_2O)$, depending on the degree of mixing of the river- and seawater. The latter behaves conservatively, that is, is only determined by the mixing ratio; $^{13}\delta$ (DIC) also depends on the DIC values of the mixing components. Therefore, the mixing line is generally not straight. The relation with $^{13}\delta$ of the bicarbonate fraction is even more complicated, as the dissociation equilibria change with pH (even the pH does not behave conservatively) (UNESCO/IAEA Volume I, chapter 9).

5.4   STABLE HYDROGEN ISOTOPES

5.4.1   The natural abundance

The chemical element hydrogen consists of two stable isotopes, ^1H and ^2H (D or Deuterium), with an abundance of about 99.985 and 0.015%, and an isotope ratio $^2H/^1H \approx 0.00015$. This isotope ratio has a natural variation of about 250‰, higher than the $^{13}\delta$ and $^{18}\delta$ variations, because of the relatively larger mass differences between the isotopes (Fig. 5.15).

As with ^{18}O, high ^2H concentrations are observed in strongly evaporated surface waters, while low ^2H contents are found in polar ice. Variations of about 250‰ are observed in components of the hydrological cycle discussed here.

5.4.2   Hydrogen isotope fractionations

The most important hydrogen isotope fractionation is that between the liquid and vapour phases of water. Under equilibrium conditions water vapour is isotopically lighter (contains less ^2H) than liquid water by amounts given in Table 5.6. Fig. 5.1 shows some actual isotope ratios for equilibrium systems and their matching ε values. The fractionation caused by diffusion of H_2O through air $(^2\varepsilon_d)$ has been measured to be $-21.5 \pm 1\%_o$, slightly more than that calculated from Eq. 2.24. Table 5.5 contains some data on the abundance and analytical aspects of $^2H/^1H$ ratios.

5.4.3   Reporting ^2H variations and the ^2H standard

VSMOW is the *standard* for $^2H/^1H$ as it is for $^{18}O/^{16}O$ ratios. No significant difference in $^2\delta$ has been detected between the original SMOW standard and VSMOW.

Reference and intercomparison water samples are available from the IAEA (Appendix II). The ^2H contents of hydrogen-containing samples are determined by completely converting

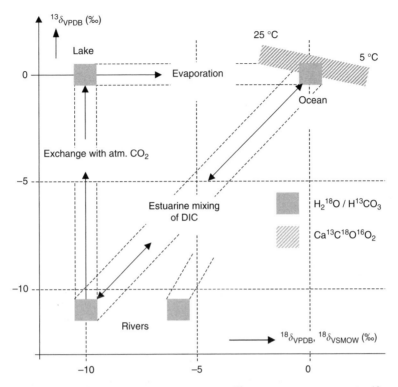

Figure 5.14.   Relation between the natural variations of $^{13}\delta$ (HCO$_3^-$ and CaCO$_3$) and $^{18}\delta$ (H$_2$O and CaCO$_3$); the graph is essentially a combination of Figs 5.4 and 5.11. Estuarine mixing only results in a straight line between the river and sea values of $^{13}\delta_{DIC}$ if C$_T$ for the components is equal. Because this is rarely the case, the relation between the two members is mostly observed as a slightly curved line. Additionally, $^{13}\delta$(HCO$_3^-$) in estuaries is not subject to conservative mixing, because the mixing process rearranges the carbonate fractions (Sections III.5.4 and 8.4.5). Depending on the residence times of water at the surface, evaporation and isotope exchange change the isotopic composition to higher δ values.

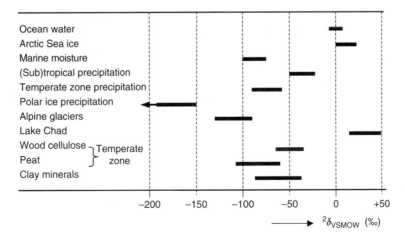

Figure 5.15.   General view of ^{2}H/^{1}H variations in natural compounds. The ranges are indicative for the majority of materials shown.

Table 5.5. The stable and radioactive isotopes of *hydrogen*: practical data for the natural abundance, properties, analytical techniques and standards. (See the Appendices for further details.)

	^1H	^2H	^3H
Natural abundance	0.99985	0.00015	$<10^{-17}$
Stability	Stable	Stable	Radioactive
Natural specific activity			<1.2 Bq/L water
Decay mode / daughter			$\beta^-/^3He$
Half-life ($T_{1/2}$)			12.320 ± 0.022 a
Decay constant (λ)			$5.576 \times 10^{-2}/a = 1/18.33$ a^{-1}
Maximum β energy			18 keV
Abundance range		250‰	0 to 10^{-16}
Reported as		$^2\delta$ or δ^2H or δD	^3A
In		‰	TU, Bq/L H$_2$O
Instrument		MS	PGC, LSS
Analytical medium		H$_2$	H$_2$O, C$_2$H$_6$, C$_6$H$_6$
Usual standard deviation		0.5‰	$\geq 1\%$ at high level
International standard		VSMOW	NBS-SRM 4361
With absolute value		$^2H/^1H = 0.00015575$	$^3H/^1H = 6600$TU or $= 0.780$ Bq/g H$_2$O as of 1 January 1988

MS = mass spectrometry, PGC = proportional gas counting, LSS = liquid scintillation spectrometry.

them to hydrogen gas. Therefore, fundamental problems of isotope fractionation during sample preparation, as with ^{18}O, do not occur. However, the analyses are more troublesome (Appendix I). More details on the measurement and calculation procedures, and on isotope reference samples are given in Appendix II.

Henceforth all $^2\delta$ values will be reported relative to VSMOW.

5.4.4   Survey of natural ^2H variations

From the foregoing it is evident that some correlation should exist between ^2H and ^{18}O fractionation effects. Hence, a relation between $^2\delta$ and $^{18}\delta$ values is to be expected in natural waters. Indeed, the $^2\delta$ and $^{18}\delta$ variations in precipitation, ice, most groundwaters and non-evaporated surface waters have appeared to be closely related. The qualitative discussion given in Section 5.2.4 for $^{18}\delta$ therefore, applies equally well to $^2\delta$. The next section is devoted to this relation.

5.5   RELATION BETWEEN ^2H AND ^{18}O VARIATIONS IN WATER

If we simply assume that evaporation and condensation in nature occur in isotopic equilibrium, the relation between the $^2\delta$ and $^{18}\delta$ values of natural waters is determined by both

Table 5.6.   Hydrogen isotope fractionation in the equilibrium system liquid water (l), water vapour (v) and ice (i); $\varepsilon_{y/x}$ represents the fractionation of compound y relative to compound x and is approximately equal to $^2\delta(y) - ^2\delta(x)$. (Values for intermediate temperatures may be obtained by linear interpolation; $T = t(°C) + 273.15K$.)

t (°C)	$^2\varepsilon_{v/l}$ (‰)[1]	$^{18}\varepsilon_{v/l}$ (‰)[2]	$^2\varepsilon_{v/l}/^{18}\varepsilon_{v/l}$
0	−101.0	−11.55	8.7
5	−94.8	−11.07	8.5^5
10	−89.0	−10.60	8.4
15	−83.5	−10.15	8.2^5
20	−78.4	−9.71	8.1
25	−73.5	−9.29	7.9
30	−68.9	−8.89	7.7^5
35	−64.6	−8.49	7.6
40	−60.6	−8.11	7.4

According to reference [1] in Table 5.4:

$$[1] \quad \ln^2\alpha = -24\,844/T^2 + 76.248/T - 0.052612 \tag{5.34a}$$

$$\text{Exponential: } ^2\alpha_{v/l} = 1.276\exp(-95.39/T) \tag{5.34b}$$

$$1/T : {}^2\varepsilon_{v/l} = -85\,626/T + 213.4‰ \tag{5.34c}$$

$$\text{According to reference [1]: } ^2\varepsilon_{i/l} = +19.3‰(\text{at } 0\,°C) \tag{5.35}$$

[2]   From Table 5.4.

equilibrium fractionations $^2\varepsilon_{v/l}$ and $^{18}\varepsilon_{v/l}$. The ratio of these factors is presented in Table 5.6 for the temperature range 0–40 °C.

A generally valid relation between the $^2\delta$ and $^{18}\delta$ values of precipitation from various parts of the world is:

$$^2\delta = 8\,^{18}\delta + 10‰ \tag{5.36}$$

This relation, shown in Fig. 5.16 and known as the *Global Meteoric Water Line* (GMWL) is characterised by a *slope* of 8 and a certain *intercept* with the ^2H axis (= the $^2\delta$ value at $^{18}\delta = 0‰$). This relation is explained quantitatively in Chapter 7.

The general relation of the MWL is:

$$^2\delta = s \times {}^{18}\delta + d \tag{5.37}$$

where the slope $s = 8$, as is explained by the ratio between the equilibrium isotope fractionations of hydrogen and oxygen for the rain condensation process (Table 5.6); d is referred to as the *deuterium excess* (*d-excess*), the intercept with the $^2\delta$-axis. In several regions of the world, as well as during certain periods of the year and even certain storms, the d-value is different from 10‰, depending on the humidity and temperature conditions in the evaporation region.

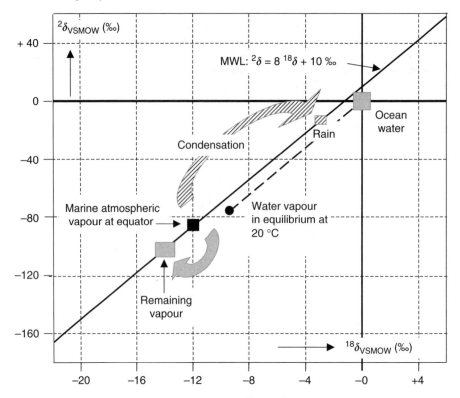

Figure 5.16.   Relation between natural variations of $^{18}\delta$ and $^2\delta$ for ocean water, atmospheric vapour and precipitation. The black round represents the hypothetical value of water vapour in isotopic equilibrium with ocean water, and the black square the observed isotopic composition of equatorial marine vapour, originated from the more realistic non-equilibrium fractionation. Marine vapour gradually condenses into precipitation (hatched arrow) with a positive fractionation, leaving the vapour progressively depleted in ^{18}O and ^{2}H (grey arrow) (cf. Fig. 5.12).

The isotopic composition of water vapour over seawater with $^2\delta = {}^{18}\delta = 0\%o$ versus VSMOW is somewhat lighter than would follow from isotopic equilibrium with the water: the evaporation is a non-equilibrium (partly kinetic) process. However, from the observed vapour composition onwards the vapour and precipitation remain in isotopic equilibrium, because the formation of precipitation is likely to occur from saturated vapour (i.e. vapour in physical equilibrium with water). Consequently, the $^{18}\delta$ and $^2\delta$ values both move along the MWL.

Conditions leading to deviations from the common MWL are mentioned in Chapter 7. For example, larger values of d are occasionally observed.

Apart from this and more importantly, deviations occur in evaporating surface waters, showing slopes of 4 to 5 rather than 8. If $^2\delta_o$ and $^{18}\delta_o$ denote the original isotopic composition of an arbitrary surface water, the δ values after evaporation are related by:

$$^2\delta - {}^2\delta_o \approx 4.5({}^{18}\delta - {}^{18}\delta_o) \quad \text{or} \quad \Delta^2\delta \approx 4.5\Delta^{18}\delta \tag{5.38}$$

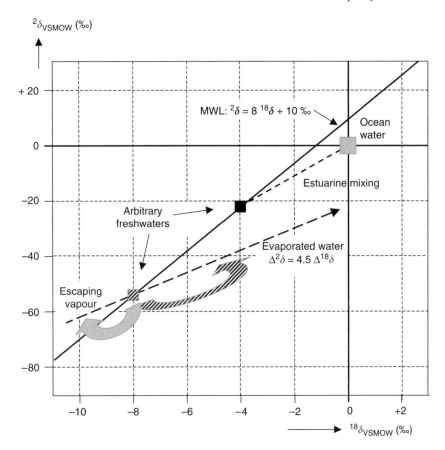

Figure 5.17. Relation between $^{18}\delta$ and $^2\delta$ for estuarine mixing and for evaporating surface water. Because the evaporation is a non-equilibrium process, isotope fractionations involved are not necessarily related by a factor of 8, as is the equilibrium condensation process, the basis for the MWL (Fig. 5.16). As in the preceding figure, the arrows indicate the direction of change of the isotopic composition of the escaping vapour and of the remaining evaporated water (evaporation line).

(Fig. 5.17). The release of relatively low-δ water vapour to the air results in an increase in δ for the remaining water, as illustrated by the model in Section 3.2.5; here for $^2\delta$ as well as $^{18}\delta$. The latitudinal temperature effect in ^{2}H of annual average precipitation results from combining Eqs 5.33 and 5.36:

$$d^2\delta/dt \approx 5.6‰/°C \tag{5.39}$$

The climatic variability of $^2\delta$ with temperature is probably also given by this relation.

CHAPTER 6

Natural abundance of the radioactive isotopes of C and H

In this chapter the two nuclides ^{14}C and ^{3}H are treated in detail, these being the most relevant radioactive isotopes in the hydrological cycle.

6.1 THE RADIOACTIVE CARBON ISOTOPE

6.1.1 Origin of ^{14}C, decay and half-life

The natural occurrence of the radioactive carbon isotope, ^{14}C or radiocarbon, was first recognised in 1946. It is naturally formed in the transitional region between the stratosphere and troposphere, about 12 km above the earth's surface, through the nuclear reaction:

$$^{14}N + n \rightarrow {}^{14}C + p \tag{6.1}$$

The thermal neutrons required are produced by reactions between very high-energy primary cosmic ray protons and molecules of the atmosphere. ^{14}C thus formed very soon oxidises to ^{14}CO, and ultimately to ^{14}CO$_2$ which mixes with the inactive atmospheric CO$_2$ (Fig. 6.1). Through exchange with oceanic dissolved carbon (primarily bicarbonate), most ^{14}CO$_2$ molecules enter the oceans and living marine organisms. Some are also assimilated by land plants, so that all living organisms, vegetable as well as animal, contain ^{14}C in concentrations about equal (Section 6.1.3) to that of atmospheric CO$_2$.

^{14}C decays according to:

$$^{14}C \rightarrow {}^{14}N + \beta^- \tag{6.2}$$

with a maximum β^- energy of 156 keV and a half-life of 5730±40 years (Fig. 6.2). The half-life was originally thought to be 5568 years, so that during the first decade or two ^{14}C age determinations were based on the wrong half-life. Later, when the better half-life became known, so many ^{14}C ages were already published that, in order to avoid confusion, it was decided that the original half-life should continue to be used for reporting ^{14}C ages. More-over, it was meanwhile known that ages would still be in error, even using the proper half-life. The fact is that during geologic times variations have occurred in the natural ^{14}C concentration of atmospheric CO$_2$ which deviate from the present. These errors were even larger.

Nowadays the *^{14}C calibration*, based on known ^{14}C content of tree rings with exactly known age, removes both errors (Section 6.1.4).

The production and distribution of ^{14}C in nature occurs through a series of chemical and biological processes which has become stationary throughout much of geologic time. As

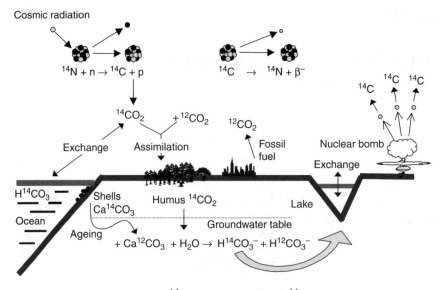

Figure 6.1.   Origin and distribution of ^{14}C in nature. The natural ^{14}C concentration by the production by neutrons from cosmic radiation has been influenced by – chronologically – the addition to the air of CO_2 without ^{14}C by the combustion of fossil fuels and the production of ^{14}C by neutrons released by the fission and fusion reactions during nuclear explosions.

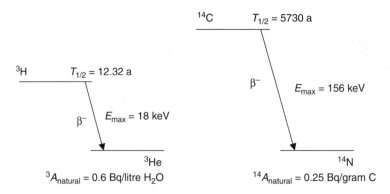

Figure 6.2.   Decay schemes for tritium and radiocarbon; both nuclides are pure β^- emitters. The present-day specific activities are given as ^3A and ^{14}A.

a consequence, the concentration of ^{14}C in the atmosphere, oceans and biosphere reached a steady-state value which has been almost constant during a geologic period which is long compared to the life span of a ^{14}C nucleus. This natural concentration, ^{14}C/C, is in the order of 10^{-12}, which is equivalent to a specific activity of about 0.25 Bq/gC (disintegrations per second per gram of carbon) (Section 4.3.3).

6.1.2   Reporting ^{14}C variations and the ^{14}C standard

Three modes of reporting ^{14}C activities are in use, in part analogous to the conventions internationally agreed (IAEA) for stable isotopes.

The *absolute (specific)* ^{14}C *activity*, that is the ^{14}C radioactivity (in Bq or, conventionally, in disintegrations per minute (dpm) per gram of carbon), is given the symbol

$$^{14}A = \text{number of disintegrations per minute (dpm/gC)} \tag{6.3}$$

It is extremely difficult to make an absolute measurement of ^{14}C activity. Moreover, the absolute ^{14}C content of a sample is generally not relevant, and the sample activities are therefore compared with the activity of a standard. In reality, the number of ^{14}C registrations (β registrations from ^{14}C decay in radiometric detectors such as proportional counters and liquid scintillation counters, or registrations of ^{14}C concentration in AMS systems) are related to the number of registrations from the standard under equivalent conditions.

This results in a ^{14}C *activity ratio* or ^{14}C *concentration ratio*:

$$^{14}a = \frac{^{14}A_{\text{sample}}}{^{14}A_{\text{reference}}} = \frac{^{14}A}{^{14}A_R} = \frac{^{14}C \text{ specific activity in the sample}}{^{14}C \text{ specific activity in the reference}}$$

$$= \frac{^{14}C \text{ concentration in the sample}}{^{14}C \text{ concentration in the reference}} \tag{6.4}$$

Because in the numerator and denominator of the last two fractions the detection efficiencies are cancelled, being equal for sample and reference material, the use of the ratio ^{14}a is adequate for any type of measuring technique.

Henceforth, the symbol ^{14}A may be used for the ^{14}C *content* (specific activity or concentration; Section 4.3.3) of a sample, whether the analytical technique applied is radiometric or mass spectrometric (AMS).

Under natural circumstances the values of ^{14}a are between 0 and 1. In order to avoid the use of decimals, it is general practice to report these values in per cent, which is equivalent to the factor 10^{-2} (see also Appendix II, Section II.5.1). It is incorrect to write $^{14}a/10^2$ instead of merely ^{14}a.

In some fields of application the differences in ^{14}C content between the project samples are small. Therefore, the practice has been adopted from the stable-isotope field of defining relative abundances, *in casu* the *relative* ^{14}C *content (activity or concentration)*, $^{14}\delta$, as the difference between sample and reference ^{14}C content as a fraction of the reference value:

$$^{14}\delta = \frac{^{14}A - ^{14}A_r}{^{14}A_r} = \frac{^{14}A}{^{14}A_r} - 1 = {^{14}a} - 1 \ (\times 10^3‰) \tag{6.5}$$

The values of δ are small numbers and therefore generally given in ‰, which is equivalent to the factor 10^{-3} (cf. Section 5.1).

In Chapter 5 we have seen that during any process in nature isotope fractionation occurs, not only for the stable isotopes, but similarly for the radioactive isotope such as ^{14}C. Neglecting this effect would introduce an error in an age determination, because the original ^{14}C content of the material would be different from what is assumed. The degree of fractionation is conveniently indicated by the $^{13}\delta$ value of the material (cf. Section 2.7):

$$^{14}\varepsilon = 2\,^{13}\varepsilon \tag{6.6}$$

An example is the apparent difference in age between C_3 plants such as trees ($^{13}\delta \approx -25‰$) and a C_4 plant such as sugar cane ($^{13}\delta \approx -10‰$), each assimilating the same atmospheric CO_2. Therefore, in defining the standard activity, but also in our treatment of all ^{14}C data, we need to *normalise* the ^{14}C results to the same $^{13}\delta$ value:

$$^{14}\delta_{\text{corrected}} - {}^{14}\delta_{\text{measured}} = 2({}^{13}\delta_{\text{corrected}} - {}^{13}\delta_{\text{measured}}) \tag{6.7}$$

By international agreement all ^{14}C results have to be corrected for a deviation of the $^{13}\delta_{\text{measured}}$ value from $^{13}\delta_{\text{corrected}} = -25‰$.

The ^{14}C *standard activity* or *concentration* was chosen to represent, as closely as possible, the ^{14}C content of carbon in naturally growing plants. The ^{14}C standard activity does not need to be, in fact it is not, equal to the ^{14}C activity of the standard. The definition of the standard ^{14}C activity is based on 95% of the specific activity of the original NBS oxalic acid (Ox1) in the year AD 1950, as discussed in more detail later. Normalisation for fractionation according to Eq. 6.7 also concerns the Oxalic Acid standard; only here the $^{13}\delta_{\text{corrected}}$ value to be applied is $-19‰$.

The absolute ^{14}C *standard activity* is defined by:

$$^{14}A^0_{\text{standard}} = 0.95\,^{14}A^0_{\text{Ox1}} = 13.56 \pm 0.07 \text{ dpm/gC} = 0.226 \pm 0.001 \text{ Bq/gC} \tag{6.8}$$

dpm/gC is disintegrations per minute per gram of carbon, while the superscript ^{0} refers to the fact that the definition is valid for the year 1950 only.

Because the original supply of oxalic acid has in most laboratories been exhausted, a new batch of oxalic acid (Ox2) is available for distribution by the NIST (formerly US-National Bureau of Standards). Through careful measurement by a number of laboratories, the ^{14}C activity has become related to that of the original Ox1 by:

$$^{14}A^0_{\text{Ox2}} = (1.2736 \pm 0.0004)\,^{14}A^0_{\text{Ox1}} \tag{6.9}$$

Both activities refer to AD 1950. Consequently, the standard activity is:

$$^{14}A^0_{\text{standard}} = 0.7459\,^{14}A^0_{\text{Ox2}} \tag{6.10}$$

where the $^{14}A^0$ values for Ox1 and Ox2 refer to the activity of the material in 1950, irrespective of the time of measurement.

Details of the measurement and calculation procedures are given in Appendix II. A summary of available ^{14}C reference samples for different compounds and ages is given in Table 11.6 in Volume I of the UNESCO/IAEA series.

6.1.3   Survey of natural ^{14}C variations

6.1.3.1   *Atmospheric CO_2*

Certain fluctuations occur in the stationary state mentioned in Section 6.1.1. Because of variations in cosmic ray intensity, and in climate affecting the size of the carbon reservoirs, the ^{14}C content of atmospheric carbon has not always been precisely the same as at present. This phenomenon has been observed by measuring the ^{14}C content of tree rings (see Fig. 6.5).

These fluctuations do not exceed a few per cent over relatively short periods, and are, therefore, of very little relevance to the hydrologist who is generally confronted with dissolved carbon from a variable vegetation covering a long and uncertain period of time.

The strong increase in the atmospheric ^{14}C level due to nuclear test explosions is of more relevance to the hydrologist. During the bomb explosions ^{14}C (and ^{3}H) is produced by the same nuclear reactions that are responsible for the natural production. In the northern hemisphere the peak concentration occurred during the spring of 1963, when it was about double the natural concentration (Fig. 6.3). In the southern hemisphere a more gradual increase has taken place. This is due to the fact that the seasonal injections of ^{14}C from the stratosphere into the troposphere primarily occurred over the northern hemisphere, while air is not easily and simply transported across the equator. The ^{14}C increase only gradually leaked through the equatorial regions to the southern troposphere.

6.1.3.2   *Vegetation and soils*

The higher ^{14}C levels have entered the soil vegetation, and have become detectable in the CO_2 gas, present in the soil and playing a major role in the formation of dissolved inorganic

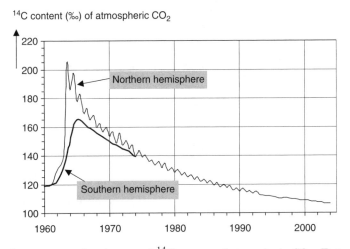

Figure 6.3.   Curve representing the natural ^{14}C content of atmospheric CO_2. Test explosions of nuclear weapons since the 1950s, and especially the early sixties, have temporarily increased the concentration by a factor of two. Due to isotopic exchange with the oceans, ^{14}C is slowly returning to 'normal'. The influence on the southern hemisphere was obviously smaller. The seasonal variation of some per cent during the late 1960s has been smoothed out, since they are not relevant in hydrology, contrary to the ^{3}H variations discussed later in this chapter (early data for Nordkapp, Norway (Nydal), later data for Groningen and Schauinsland, Southern Germany (Levin and Kromer, 2004); data representative for the Northern hemisphere; see References and selected papers).

carbon in groundwater. In this way the 'bomb effect' possibly increases the ^{14}C content of young groundwater.

The CO_2 generated in the soil by decaying plant remains and by root respiration is relatively young, and therefore contains about the atmospheric ^{14}C concentration ($^{14}a =$ 100%) (Fig. 6.3).

6.1.3.3   *Seawater and marine carbonate*

Because seawater is in constant exchange with atmospheric CO_2, one would expect that, like the stable carbon isotopes, isotopic equilibrium exists between the two reservoirs. This is not the case, due to the phenomenon of *upwelling* of water from greater depth and a considerable age. This may extend to ages of 1500 years, equivalent to an ^{14}a value in the order of 85%. In general, surface seawater ^{14}C contents are observed to approximate 95%.

6.1.3.4   *Groundwater*

By processes of erosion and re-sedimentation, fossil carbonate is generally part of terrestrial soils. Here it may be dissolved by the action of soil CO_2, which is contained in the infiltrating rain water. In this way the dissolved inorganic carbon in groundwater also contains ^{14}C (Fig. 6.4) As the carbonate is very old and thus without ^{14}C ($^{14}a = 0\%$) – generally, but not always (see Section 9.4 and UNESCO/IAEA Volume IV on Groundwater) – the bicarbonate resulting from the reaction:

$$CO_2 + H_2O + CaCO_3 \rightarrow Ca^{2+} + 2HCO_3^- \qquad (6.11)$$

will have a ^{14}C content one half that of the CO_2 ($^{14}a = 50\%$) (cf. ^{13}C, Fig. 5.6). Isotopic exchange with soil CO_2 or atmospheric CO_2 lead to higher ^{14}C concentrations for the

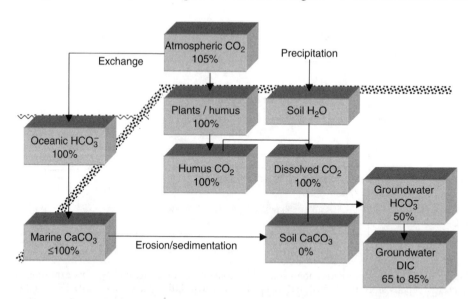

Figure 6.4.   Schematic representation of the formation of dissolved inorganic carbon in groundwater. The percentages are the ^{14}a values indicative for the carbonate dissolution, which is the most relevant process in most types of groundwater. The ^{14}a value of groundwater HCO_3^- of 50% results in higher ^{14}a values in the total dissolved carbon because of the additional CO_2 and isotopic exchange with the soil CO_2.

inorganic carbon fraction of recent groundwater, in combination with respectively decreased or increased $^{13}\delta$ (see Section 9.4). The increased level of ^{14}C in atmospheric CO_2 since 1963 can lead to ^{14}C contents in soil organic matter and soil CO_2 which exceed the natural atmospheric value.

6.1.4 ^{14}C age determination

6.1.4.1   *Dating terrestrial samples*
Knowing the rate of radioactive decay (λ or $T_{1/2}$), the age (T = time elapsed since death) of a carbonaceous sample, organic or inorganic, can be calculated from the measured activity, ^{14}A, if the ^{14}C activity at the time of death, $^{14}A^{initial}$ is known (Eq. 4.21):

$$T = -(T_{1/2}/\ln 2)(^{14}A/^{14}A^{initial})$$

Considering this equation, three quantities have to be established: $T_{1/2}$, $^{14}A^i$ and the proper ^{14}A value of the sample ^{14}C activity. As explained before, the condition that sample and reference activity are determined at the same time and under similar conditions results in an ^{14}a value valid for the year 1950; that is the age obtained counts back from 1950.

By international convention:

1. The initial activity equals the *chosen* standard activity in AD 1950.
2. The ^{14}C activities have to be normalised for fractionation (samples to $^{13}\delta = -25‰$, Ox1 to $-19‰$, Ox2 to $-25‰$) (see Appendix II).
3. The original ('Libby') half-life of 5568 has to be used.

This results in the relation:

$$T = -\frac{T_{1/2}}{\ln 2} \ln \frac{^{14}A_{sample}}{^{14}A_{standard}} = -\frac{5568}{0.693} \ln \frac{^{14}A^0_{sample}}{^{14}A^0_{standard}} = -8033 \ln {}^{14}a_{sample} \qquad (6.12)$$

The result is called the *conventional* ^{14}C age BP of the sample that is, the age Before Present (BP) (by definition: before AD 1950).

> In order to avoid confusion, the original (conventional) half-life of $T_{1/2} = 5568$ years is still being used by international agreement for geological and archaeological dating. In hydrology, however, we apply the true half-life of 5730 years.

The practice using these conventional ages is neat but of course a simplification. Apart from using the 'wrong' half-life, the assumption of a constant ^{14}C content for living organic matter is proven untrue. However, by applying the ^{14}C *calibration curve*, that is, the empirical relation between the conventional age of tree rings (to 10 000 years BC) and their real age, both problems are solved and ^{14}C ages are translated into real ages (Fig. 6.5). The 'wiggles' of the calibration curve are due to natural variations in the atmospheric ^{14}C content, as mentioned in Section 6.1.3. Further details about radiocarbon dating are given in the list of references.

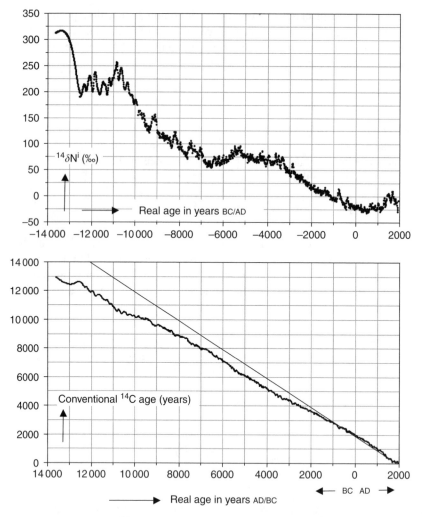

Figure 6.5.   Variations in the ^{14}C content of atmospheric CO_2 in the past (upper curve), resulting from ^{14}C measurements of tree rings of accurately known age by a number of laboratories, and of annual growth rings in corals. The parameter showing the deviations from the standard is defined by Eq. II.27 in Appendix II. The data result in the calibration curve (lower curve), used for translating conventional ^{14}C ages into real ages. The straight line in the calibration curve represents $y = x$, that is, if there would be no difference between the ^{14}C age and the real age. The graph shows that at high ages there is a considerable difference between the ^{14}C age and the true age.

6.1.4.2   *Dating groundwater*

Dating water as such has no meaning. Dating groundwater determines the age of groundwater, that is, the time elapsed since the water became groundwater, in other words, since it infiltrated into the soil as precipitation or any other type of surface water (rivers, lakes).

In principle there are some methods for calculating or estimating groundwater ages, for instance, based on hydrodynamical modelling. Here we discuss the application of radioactive decay.

Essential to know is the input function, that is, the concentration or the specific activity of the radioactive tracer that infiltrated. The water molecule itself has one radioactive isotope, tritium or ^3H. The disadvantages are that both the activity and the half-life are very small. Possible applications are discussed in Chapter 9. Therefore, we need to revert to radioactive compounds dissolved in the water, rather than the water itself, such as ^{14}C, and Uranium isotopes (the latter in Chapter 12 of Volume I of the UNESCO/IAEA series). ^{14}C has a very attractive half-life for solving hydrological problems. It is contained by DIC, as well as dissolved organic molecules (DOC).

If $^{14}a^i$ were known and if no carbon entered or escaped the groundwater in the course of time, the water age (since infiltration) would be given by:

$$T = -8267 \ln[(^{14}A/^{14}A^0_{standard})/(^{14}A^i/^{14}A^0_{standard})] = -8267 \ln(^{14}a/^{14}a^i)$$

(6.13)

It is tempting to apply this ^{14}C method to DIC, since the sampling and measuring techniques are relatively simple and available.

However, we warn the reader that, for geochemical reasons, to obtain trustworthy groundwater ages by applying the radiocarbon dating technique to DIC is by no means simple and straightforward.

Volume IV of the UNESCO/IAEA series discusses the various problems. Here we present a brief review. More details are given in Chapter 9.

The initial ^{14}C activity $^{14}A^{initial}$ $(=^{14} A^i)$ in the total DIC content of infiltrating groundwater cannot simply be taken as being equal to $^{14}A_{standard}$, as in the procedure for calculating conventional ages of terrestrial carbon (Eq. 4.21). To estimate the original, real value $^{14}A^i$ or $^{14}a^i$ of recent groundwater, that is, the ^{14}C content of DIC during the formation/infiltration, is more complicated as the chemistry, possibly even changing in time, plays an essential role. Some important data on the inorganic carbon chemistry of water are given in Appendix III.

In general, the formation of DIC in groundwater essentially comprises two processes: (i) the dissolution of carbonate by H^+ in H_2CO_3 from soil CO_2 or from atmospheric CO_2 (in arid regions) and (ii) the dissolution of additional CO_2 related to the pH of the water. Various models deal with the development of DIC and the concurrent $^{13}\delta$ and ^{14}a values.

The CO_2–$CaCO_3$ concept is based on the mass balance of inorganic carbon (Fig. 6.4). Rewriting Eq. 6.11 in a more complete form gives:

$$(a + 0.5b)\, CO_2 + 0.5b\, CaCO_3 + H_2O \rightarrow 0.5b\, Ca^{2+} + b\, HCO_3^- + a\, CO_2$$

(6.14)

where a and b are the respective concentrations of dissolved carbon dioxide and bicarbonate. In this simple concept the chemical composition of the water, *in casu* the carbonic acid fractions, determine the original ^{14}a value for the mixture of CO_2 and HCO_3^-:

$$(a + b)^{14}a = 0.5b^{14}a_l + (a + 0.5b)^{14}a_g$$

(6.15)

where the subscripts l and g refer to the solid carbonate and the gaseous CO_2, respectively. As in most groundwater, pH is low so that the only carbonic acid fractions are the dissolved CO_2 (a) and HCO_3^- (b). With the ^{14}a values given in Fig. 6.4, the dissolved bicarbonate is expected to have a ^{14}C content of 50%, while an additional amount of CO_2 (with $^{14}a = 100\%$) shifts the ^{14}C content of DIC ($= a + b$) to a somewhat higher value. In temperate climates where the land has a vegetation cover, the observed values of $^{13}\delta_{DIC}$ and $^{14}a_{DIC}$ are generally in the range of -11 to $-14‰$ and 65 to 85%, respectively. The various procedures for estimating the $^{14}a^{initial}$ value for the DIC content of groundwater are discussed in more detail in Chapter 9.

6.2   RELATION BETWEEN ^{13}C AND ^{14}C VARIATIONS

A summary of the ^{14}C (^{14}a in %) and ^{13}C content ($^{13}\delta$ in ‰) values in reservoirs relevant to the hydrological cycle is given in Fig. 6.6.

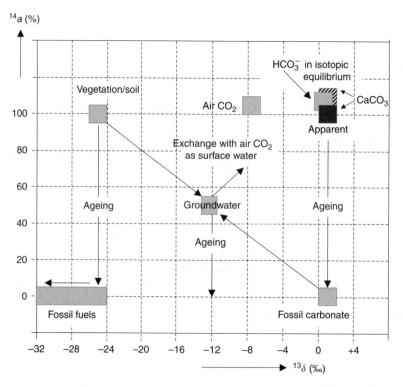

Figure 6.6.   Schematic representation of the relation between variations in ^{13}C and ^{14}C in nature. The groundwater values have been explained in the Figs 5.6 and 6.4. The marine bicarbonate and solid carbonate (vertically dashed) values for the case of an isotopic equilibrium with atmospheric CO_2 are not observed in the ocean. The upwelling of deep, relatively old (1500 a) water causes the surface water to be lightly depleted in ^{14}C. The apparent age of this water – and consequently of the marine carbonate formed in this water (black square) – is about 400 years (equivalent to 5% ^{14}C).

The values are merely indicative and do not exclude deviations. The figure is a combination of the discussions represented schematically in Figs 5.4, 5.6 and 6.4.

Special attention needs to be given to the value for marine carbonate in Fig. 6.6. By coincidence, the observed ^{14}a values for vegetation and marine carbonate are almost equal. The fact is that two processes cancel each other: (i) starting with organic matter on land with $^{13}\delta$ about $-25‰$ and $^{14}a = 100\%$ (by definition, cf. Section 6.1.2), the difference in $^{13}\delta$ between land vegetation and marine carbonate (via atmospheric CO_2) of about $25‰$ requires ^{14}a of the latter to be 5% ($= 2 \times 25‰$) larger; (ii) the upwelling of deep sea water with ages of up to 1500 years generally causes the surface ocean water to be depleted by 5%.

6.3   THE RADIOACTIVE HYDROGEN ISOTOPE

6.3.1   Origin of ^{3}H, decay and half-life

The radioactive isotope of hydrogen, ^{3}H (tritium or T), originates (as does ^{14}C) from a nuclear reaction between atmospheric nitrogen and thermal neutrons:

$$^{14}N + n \rightarrow {}^{12}C + {}^3H \tag{6.16}$$

The ^{3}H thus formed enters the hydrologic cycle after oxidation to $^1H^3HO$ (Fig. 6.7). It finally decays according to:

$$^3H \rightarrow {}^3He + \beta^- \tag{6.17}$$

with $E_{\beta\,max} = 18$ keV. The most recent value of the half-life is 12.32 ± 0.02 year (Fig. 6.2).

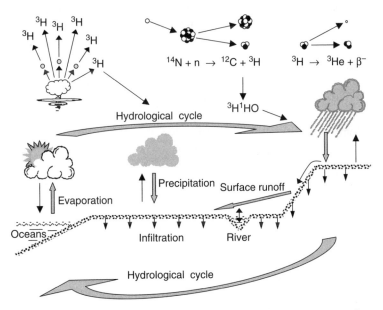

Figure 6.7.   Origin and distribution of ^{3}H in nature. Contrary to ^{14}C, the turnover of ^{3}H is very fast, except where it is fixed in glacier ice or groundwater.

6.3.2   Reporting ^{3}H activities and the ^{3}H standard

Because the nuclear reaction (Eq. 6.16) has a lower probability than reaction (Eq. 6.1) and because the residence time of ^{3}H in the atmosphere is much smaller than that of ^{14}C, the natural ^{3}H concentration in the air is much smaller than that of ^{14}C. Natural ^{3}H abundances are either presented as specific activities (in Bq per litre of water) or in Tritium Units (TU), the latter by definition equivalent to a concentration of ^{3}H/^{1}H $= 10^{-18}$ (1 TU $= 3.21$ pCi/L $= 0.119$ Bq/L) (Section 4.3.3).

For ^{3}H, as for ^{14}C, it is extremely difficult to determine the absolute specific activities. Tritium activities are therefore related to a reference water sample which is measured under equivalent conditions. For this purpose the IAEA and NIST (the former NBS) have available a tritium standard NBS-SRM 4361 of 11 100 TU as of 3 September 1978. After being related to this standard, ^{3}H concentrations are reported in absolute values (^3A in TU), corrected for decay to the date of sample collection.

6.3.3   Survey of natural ^{3}H variations

Under undisturbed natural conditions the ^{3}H concentration in precipitation was probably about 5 TU, which is equivalent to a specific activity of about 0.6 Bq/L.

Following the nuclear weapon tests of the early 1960s, the ^{3}H content of precipitation temporarily increased by a 1000-fold in the Northern hemisphere (Fig. 6.8). Since 1963 this extreme ^{3}H content has decreased to essentially natural values in winter and about twice natural in summer.

A large part of the ^{3}H (as well as ^{14}C) produced by the nuclear explosions has been injected into the stratosphere and returns to the troposphere each year during spring and early summer. This causes the seasonal variation in both ^{3}H and ^{14}C, being more pronounced in the former, because the residence time of H_2O to which ^{3}H is coupled in the atmosphere is very small (in the order of weeks). The probability of contamination of young groundwater

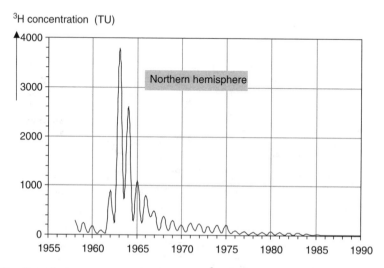

Figure 6.8.   Smoothed curve representing the average ^{3}H content of precipitation over the continental surface of the Northern hemisphere at the time of sample collection. The more recent values of ^{3}H in precipitation scatter between 5 and 10 TU, and are thus almost back to natural.

by bomb ^{3}H prevents the water being simply dated by measuring the degree of ^{3}H decay. Nevertheless, ^{3}H data can often be used to determine dates *ante quem* or *post quem*. For example, water with $^3A < 5$ TU must have a mean residence time of more than 40 years; water having $^3A > 20$ TU must date from after 1961. The tracer use of ^{3}H is further discussed in Volume IV of the UNESCO/IAEA series and in Chapter 9.

6.4   COMPARISON OF ^{3}H AND ^{14}C VARIATIONS

6.4.1   Relation between ^{3}H and ^{14}C in the atmosphere

We first compare the curves showing ^{14}C and ^{3}H variations in atmospheric CO_2 and H_2O, respectively (Figs 6.3 and 6.8), because they give important information on the global carbon and water cycle. The main differences are the following:

1. The effect of the nuclear bomb ^{14}C and ^{3}H had a much more pronounced effect on the atmospheric content of the latter, because the natural ^{3}H concentration is so much lower.
2. The restoration to natural conditions proceeds faster for ^{3}H than for ^{14}C, because the turnover time for water in the atmosphere from evaporation of ocean water to precipitation is very short (in the order of weeks), while the exchange of CO_2 between the atmosphere and ocean water is very slow (in the order of years).
3. For the same reason, ^{14}C shows much smaller seasonal amplitudes than ^{3}H; the consequence for the latter is that in temperate climates, where the main infiltration of rain occurs during the winter half-year, groundwater rarely shows the high ^{3}H values observed in precipitation during the early 1960s.
4. The post-1960 ^{14}C increase in the air over the Southern hemisphere is less pronounced than over the Northern hemisphere, as the majority of nuclear tests took place in the north, and the air is not easily transported across the equator. For ^{3}H the consequences are that the equivalent curve for the Southern hemisphere is much different; it is less regular, shows a maximum at 30 TU around 1963–65 with occasional peaks to 80 TU, and a slow decrease to the natural level (5 TU) at present.
5. ^{14}C is being used to calibrate global carbon cycle models, indicating the exchange (and transition) time of (additional) CO_2 between air and sea; ^{3}H is being applied in oceanography, indicating the rate of vertical water movement and lateral water flow.

6.4.2   Relation between ^{3}H and ^{14}C in groundwater

We have already emphasised that our discussion of the occurrence of ^{14}C and ^{3}H in groundwater may not lead to the principle that, using these isotopes, we are able to simply measure the *age of groundwater*; that is, the period of time elapsed since the water infiltrated.

The main obstacle is the unknown isotopic values at time zero, the moment of infiltration, for ^{3}H as well as for ^{14}C. However, ^{14}C and ^{3}H together certainly offer the possibility to set limits to *absolute ages*, especially in combination with hydrogeological and hydrochemical evidence.

Determining *relative ages* (i.e. age differences between neighbouring samples, which is often equally relevant) is less complicated, at least if the chemistry does not indicate the action of processes underground – such as carbonate dissolution, decomposition of

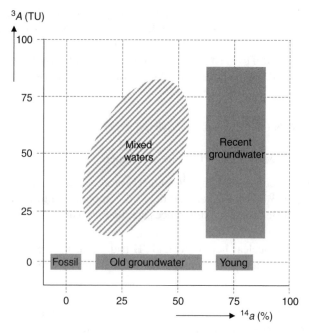

Figure 6.9.   Schematic representation of relations between the ^{14}C and ^{3}H content in groundwater. The scheme is merely indicative and serves to give a general impression. The terminology is discussed in the text.

additional (old) organic matter and certainly if we are dealing with a mixture of different water – that could influence the carbon isotopic composition. The simple case of the age difference between water samples collected at two geographical locations k and k+1 is expressed as:

$$\Delta T = 8270 \ln\frac{^{14}a_{k+1}}{^{14}a_k} \tag{6.18}$$

where $(\ln 2)/T_{1/2} = 8270$ years. From a qualitative point of view, several approximate conclusions can be drawn from isotopic data. For example, if the water sample shows a measurable ^{3}H activity (larger than 1 TU), the water is *subrecent*; that is, less than 50 years old, or is a mixture of young and old water, or at least contains an admixture of a certain percentage of recent water. With no measurable ^{14}C and ^{3}H the water is certainly a few tens of thousands of years old. If the water is very recent, the ^{14}C content is expected to be close to or above 100%, because the soil CO_2 is then likely to contain bomb ^{14}C.

Volume IV of the UNESCO/IAEA series is especially devoted to these problems and (im)possibilities.

A general and simplified summary is shown in Fig. 6.9. With *recent groundwater* we refer to water that infiltrated not more than some tens of years ago. *Young groundwater* may have any age of hundreds of years, and old groundwater thousands. Finally, *very old*, also referred to as *fossil groundwater*, does not contain ^{3}H and ^{14}C and is several tens of thousands of years old. Waters which are relatively high in ^{3}H and lower in ^{14}C than recent water are likely to be mixtures of young and old water.

CHAPTER 7

Isotopes in precipitation

This chapter deals with the distribution of natural isotopes in precipitation. The radioactive isotope of hydrogen, ^3H, serves to trace water masses and to look for their origin. The stable isotopes of oxygen and hydrogen, ^{18}O and ^2H, are applied for tracing water in its various stages in migration over and in the soil, and for studying evaporation. ^{14}C enters the dissolved carbon cycle and plays an important role in determining the 'age' of groundwater together with ^{13}C. The carbon isotopes are discussed further with surface- and groundwater aspects.

7.1 ^{18}O AND ^2H IN PRECIPITATION

In this section we discuss the isotopic composition of the various stages in the precipitation process in a kind of natural order. Water is evaporating from the sea; the marine vapour for a large part precipitates over the oceans as it is transported to higher latitudes and altitudes, where the vapour cools and condenses. Part of the vapour is brought to the continents where it precipitates and forms different modes of surface- and groundwater. The 'last' marine vapour is precipitated as ice over the Arctic and the Antarctic.

Compared to the ocean water, the *meteoric waters* (i.e. the atmospheric moisture, and the precipitation and ground- and surface water derived from them) are mostly depleted in the heavy isotopic species: ^{18}O, ^{17}O and ^2H. The main reason for depleted values of meteoric water is the Rayleigh rainout effect, which operates on a limited water (vapour) reservoir in the atmosphere, and progressively removes rain from the vapour mass that is enriched in the heavy isotopes relative to the vapour. The average ocean composition is accepted as the reference standard for these isotopes (Section 2.8) so that δ (Standard *M*ean *O*cean *W*ater) = 0‰ by definition

All $^2\delta$ and $^{18}\delta$ values of water are given relative to the VSMOW standard as $\delta_{VSMOW}(x)$.

Most δ values of meteoric waters are negative. A low value is $^{18}\delta$ of Antarctic ice: -50‰.

Our picture of the global distribution of isotopes in meteoric waters is derived from the data of the Global Network of Isotopes in Precipitation (GNIP), established by the International Atomic Energy Agency (IAEA) in co-operation with WMO in 1961 (see Box). In this programme, monthly pooled samples of precipitation are collected world-wide and analysed for their ^{18}O, ^2H and ^3H contents. The annually averaged $^{18}\delta$ values

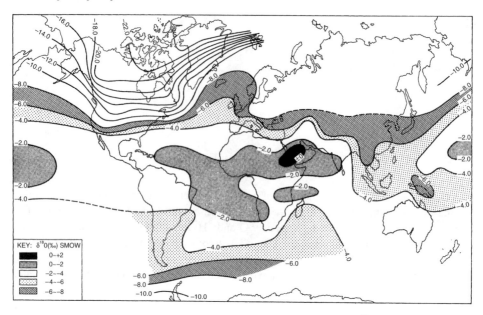

Figure 7.1.   World-wide distribution of the annual mean of precipitation $^{18}\delta$, based on the GNIP data set (UNESCO/IAEA Volume II).

are shown in Fig. 7.1. The degree of depletion is related to geographic parameters such as latitude, altitude and distance from the coast, and to the fraction precipitated from a vapour mass. Each of these aspects is discussed in more detail later in this chapter.

While the monthly sampling regime can serve to specify the inputs to larger hydrological systems, more detailed sampling may be required for regional water studies. Furthermore, more detailed data on precipitation and atmospheric moisture will be necessary in order to understand the effect of changes in climate and of the surface/atmosphere interaction pattern on the isotopic signal of these lumped monthly data.

The Global Network of Isotopes in Precipitation (GNIP)

In the early 1960s a precipitation sampling network was established by IAEA in Vienna and the World Meteorological Organisation (WMO) in Geneva, with a view to document the isotopic parameters $^2H/^1H$, $^{18}O/^{16}O$ and ^3H, together with other meteorological parameters of the input into hydrological systems. The network consisted of about 100 sampling sites world-wide, including marine, coastal and inland stations. Samples are still being collected monthly and analysed, although the network has been slightly reduced and modified over the years. Some local and regional networks and stations were also added over shorter periods of time.

The relevant data have been published regularly in the Technical Report Series of the IAEA, but have lately become available on Internet as *GNIP Data* (see http://isohis.iaea.org; it is advised to download per WMO Region).

7.1.1   Evaporation of seawater and the marine atmosphere

The source of water in the atmosphere is the evaporation of water on the surface of the Earth, foremost from the oceans and open water bodies. To a lesser extent, evaporation from plants (referred to as transpiration) and from the soil adds to the evaporation flux into the atmosphere. The isotope fractionation which accompanies the evaporation process is one important factor in the variability of isotopic composition within the water cycle. In the first instance it determines the isotopic composition of the oceanic water vapour mass from which the rain precipitates.

Observations – confirmed by model calculations – show that the range of δ values for atmospheric vapour over the vapour source regions, *in casu* mainly the tropical oceans, are:

$$^{18}\delta \approx -12 \pm 1\%_0$$
$$^{2}\delta \approx -85 \pm 5\%_0$$

$$(7.1)$$

7.1.2   Formation of precipitation

The formation of precipitation comes about as a result of the lifting of an air mass (dynamically or orographically). Due to adiabatic expansion, the air mass cools until the dew point is reached. Provided appropriate condensation nuclei are present, cloud droplets are formed. These are believed to be in local isotopic equilibrium with the moisture in the warm part of the cloud, due to a rapid exchange which takes place between the droplets and the air moisture (see Gat: Volume II of the UNESCO/IAEA series).

The exchange of water molecules between a liquid drop and the ambient water vapour results in the establishment of isotopic equilibrium in those cases where the air is saturated with respect to the liquid at the prevailing temperature. Evaporation or condensation of water occurs if the air is respectively under- or over-saturated with regard to the saturated vapour pressure, accompanied by isotope fractionation characteristic of these processes. Once equilibrium has been established, a dynamic exchange of water molecules continues without leading to a visible change (stationary state between vapour and droplets).

Where the ambient air is under-saturated, evaporation of the droplets occurs. As a result there is an isotopic enrichment of the water in the liquid phase, usually along ($^{18}\delta, ^{2}\delta$) evaporation lines with slopes smaller than 8 (see Section 8.1.3).

7.1.2.1   *Temperature dependence of $^{18}O/^{16}O$ and $^{2}H/^{1}H$ in precipitation*
The further one moves from the water source, the more depleted in heavy isotopes are the meteoric waters, resulting in the so-called *altitude, latitude* and *continental* effects. All these effects are basically related to the wringing out of moisture by cooling the air masses, and indeed the correlation with temperature appears as the overriding factor.

As the air mass cools, precipitation is formed in isotopic equilibrium with the vapour. At thermodynamic equilibrium between vapour and water the latter has a higher ^{18}O and ^{2}H content. Thus, the remaining vapour is continuously and progressively depleted in the heavy isotope. Fig. 7.2 presents the simple box model for this process. N_V is the number of water molecules, which closely equals that of the abundant, isotopically lighter molecules; R_V is the ratio of the isotopic molecules for either $^{2}H/^{1}H$ or $^{18}O/^{16}O$.

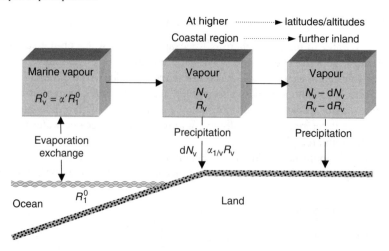

Figure 7.2.   Schematic representation of a simplified (non-recycling) Rayleigh model applied to evaporation from the ocean (with non-equilibrium fractionation α') and global precipitation. Water vapour originating from oceanic regions with strong evaporation moves to higher latitudes and/or altitudes with lower temperatures. The vapour gradually condenses to precipitation and loses $H_2{}^{18}O$ more rapidly than $H_2{}^{16}O$, because of isotope fractionation, causing the remaining vapour and also the 'later' precipitation to become more and more depleted in both ^{18}O and ^2H. Determination of the effective fractionation α' is complicated theoretically. δ_v^0 values are given in Section 7.1.1.

When dN_V molecules are removed with an accompanying fractionation factor $\alpha_{l/v}\ (=1/\alpha_{v/l})$, the rare isotopic mass balance for the transported vapour is now written as:

$$R_v N_v = (R_v - dR_v)(N_v - dN_v) + (\alpha_{l/v} R_v)dN_v$$

resulting in:

$$dR_v/R_v = (\alpha_{l/v} - 1)(dN_v/N_v)$$

and, if we assume that the number of isotopically light molecules, N_v, equals the total amount of molecules, the solution for this equation is:

$$R_v/R_v^0 = (N_v/N_v^0)^{\alpha_{l/v}-1} \tag{7.2}$$

where the superscript 0 refers to the initial conditions, that is, the source region of the water vapour.

For isobaric cooling the relative change in the amount of vapour is taken equal to the relative change in the saturated vapour pressure p_V:

$$dN_v/N_v = dp_V/p_V$$

Further, the relation between vapour pressure and temperature for an isobaric condensation process is presented by the law of Clausius Clapeyron:

$$p_V = C \exp(-D/T) \tag{7.3}$$

where

T = absolute temperature = t (°C) + 273.15 K

D = L/G = 5349 K

L = molar heat of evaporation = 44.4 × 10^3 J/mole

G = gas constant = 8.3 J/Kmole

C = constant for water (value here irrelevant)

From this we have:

$$dN_v/N_v = (D/T^2)dT$$

or

$$\frac{N_v}{N_v^0} = \frac{p_v}{p_v^0} = e^{D(1/T_0 - 1/T)}$$

so that

$$\frac{R_v}{R_v^0} = e^{-D(1/T - 1/T_0)(\alpha_{l/v} - 1)} \tag{7.4}$$

where T refers to the temperature of the sampling station, and T_0 to the source region of the water vapour.

For the temperature dependence of the fractionation between water vapour and liquid we choose the exponential equation ($\alpha_{l/v} = 1/\alpha_{v/l}$ in Table 5.4):

$$^{18}\alpha_{l/v} = Ae^{B/T} = 0.9845\, e^{7.430/T} \tag{7.5}$$

The isotopic ratio of precipitation condensing from atmospheric water vapour is: $R_l = \alpha_{l/v}R_v$, where $\alpha_{l/v}$ is determined by the condensation temperature, or rather the temperature at the cloud base. Combining Eqs 7.4, 7.5 and the value for $R_v^0 = R_{VSMOW}(1 + {}^{18}\delta_v^0)$ with $^{18}\delta_v^0 = -12‰$ (Sectioin 7.1.1), and considering that $R_l^0 = R_{VSMOW}$, leads to:

$$\delta_{\text{prec. at temp } t} = \delta_l = \alpha_{l/v}\frac{R_v}{R_v^0}\frac{R_v^0}{R_l^0} - 1 \tag{7.6}$$

This equation permits us to calculate curves of $^{18}\delta$ values versus temperature (Fig. 7.3), which may be compared with the observed variations in average annual $^{18}\delta$ values with latitude. Of course the model is too simple, as it is assumed that all vapour originates from regions around the thermal equator and all ocean evaporation at higher latitudes is neglected.

Using the given equations and assuming the temperature t_0 in the vapour source region to be 25 °C, the calculated temperature dependence of the latitude effects for $^{18}O^{16}O$ and $^2H/^1H$ are in the range of

$$d^{18}\delta/dt = 0.85 - 0.017t \ (‰/°C) \tag{7.7}$$

(from +0.8‰/°C at 0°C to +0.5‰/°C at 20°C)

$$d^2\delta/dt = 7.6 - 0.17t \ (‰/°C) \tag{7.8}$$

(from +7.6‰/°C at 0°C to +4.2‰/°C at 20°C)

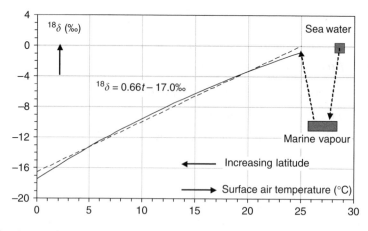

Figure 7.3. Result of calculating the temperature dependence of the latitude effect according to the simplified model of Figs 5.12 and 7.2, using Eqs 7.4, 7.5 and 7.6, with $^{18}\delta$ of the original marine vapour $-12‰$ (Eq. 7.1), and the evaporation temperature 25 °C. The right-hand side of the drawing refers to the formation of water vapour from seawater and the subsequent condensation to form precipitation. The equation refers to the linear best fit (dashed line) for the calculated result.

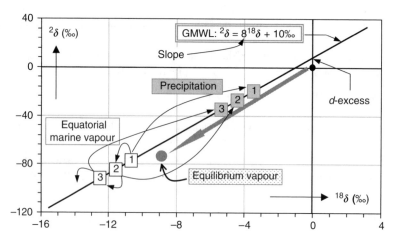

Figure 7.4. Schematic representation of the isotopic consequences of (non-equilibrium) evaporation from the oceans (black slice at (0,0)) forming the marine atmospheric vapour (white squares). Hypothetical equilibrium fractionation (grey arrow) would have resulted in a smaller fractionation (grey slice). The figure furthermore shows the progressive depletion of the vapour mass and thus of the precipitation (light squares) by the (here stepwise) condensation process, which preferentially removes the ^{18}O and ^{2}H isotopes from the vapour.

7.1.2.2 *Relation between $^{18}O/^{16}O$ and $^{2}H/^{1}H$ in natural waters*

The variations in $^{18}\delta$ and $^{2}\delta$ can often be better understood if we consider the combination of the stable isotopes of hydrogen and oxygen.

These processes and the resulting combination of ^{18}O and ^{2}H isotope effects are stepwise (and thus unrealistically) visualised in Fig. 7.4. The evaporation of seawater is in part a non-equilibrium process. This results from the fact that the air above the sea surface is under-saturated with respect to water vapour, and that the rate-determining step is one of diffusion from the surface to the marine air. If it were saturated, the isotopic composition ($^{18}\delta$, $^{2}\delta$) would move along the grey arrow, as determined by the equilibrium fractionations for ^{18}O and ^{2}H (Table 5.6).

Once the (non-isotopic-equilibrium) vapour has been formed (open square no. 1), the rainout process proceeds in isotopic equilibrium, as the vapour is then saturated. Removing the 'first' rain (grey square no. 1) causes the remaining vapour (no. 2) to be depleted in both isotopes. This process continues: vapour and condensate (= rain) become progressively depleted, and the isotopic compositions 'move' along a meteoric water line (black line in Fig. 7.4), the slope of which is given by the ratio of $^{2}\varepsilon_{v/l}/^{18}\varepsilon_{v/l}$ (see Table 5.6).

Eqs 7.7 and 7.8 indicate the relations between ^{18}O/^{16}O and ^{2}H/^{1}H for precipitation and the air temperature, resulting in the MWL:

$$^{2}\delta = (8.4 \pm 0.4)\,^{18}\delta + (11.5 \pm 2.0)\%_o \tag{7.9}$$

The *slope* (≈ 8) is quite consistent for non-evaporating meteoric waters; the *deuterium-excess* (*DXS*) ($\approx 10\%_o$) primarily depends on the relative humidity in the vapour source region.

For (surface) water (or rain droplets) subject to evaporation the conditions concerning the ($^{18}\delta$, $^{2}\delta$) relation are such that the slopes of the evaporation lines are generally different from 8. Consequences of evaporation are discussed further in Chapter 8 on Surface Waters.

The GNIP data pertain to precipitation samples. The atmospheric vapour is always much more depleted in the heavy isotopic species, by close to $10-12\%_o$ in $^{18}\delta$ on average. In a continental setting in temperate and humid regions, the air moisture and precipitation are found to be close to isotopic equilibrium with each other at the prevailing temperature. On the contrary, this is not strictly true in the vapour source regions, that is, in the oceanic setting, as well as over evaporating surface water bodies, but here the physical conditions are different. The atmosphere contains under-saturated vapour, otherwise there would be no evaporation, while during precipitation the air has to contain saturated vapour or otherwise no condensation can take place.

7.1.3 Observed stable isotope effects in precipitation

This section discusses data relating to the isotopic composition of precipitation. In principle, variations in $^{18}\delta$ and $^{2}\delta$ are coupled one way or another: under conditions of isotopic (and thus also physical) equilibrium, according to the meteoric water line with slope 8 (the GMWL), and if the conditions are non-equilibrium by a more complicated kinetic process. For less detailed hydrological surveys it is generally assumed that $^{18}\delta$ and $^{2}\delta$ values are coupled as if in equilibrium. In the next sections we therefore also report on data from literature that are concerned with only $^{18}\delta$ or $^{2}\delta$.

> However, it should be emphasised that measuring the oxygen as well as hydrogen isotopes often presents additional information; the combination of $^{18}\delta$ and $^{2}\delta$ is to be preferred.

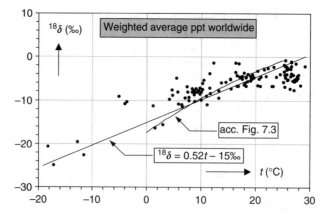

Figure 7.5. The *latitude or annual-temperature effect* on $^{18}\delta$ of precipitation (ppt): weighted $^{18}\delta$ of total precipitation over periods of at least one decade from marine/coastal, continental and (Ant)Arctic stations are shown as a function of the mean measured surface air temperature (data from the GNIP network). The $(t, {}^{18}\delta)$ curve according to Fig. 7.3 and that according to Eq. 7.11 are given for reference.

7.1.3.1 *The latitude/annual-temperature effect*

To explain in brief and numerically the stable isotopic composition of precipitation on a global scale, we have applied the Rayleigh model in Section 7.1.2.1. The progressive rain-out process, based on the Rayleigh fractionation/condensation model, results in a relation between the observed annually averaged $^{18}\delta$ and $^{2}\delta$ values for the precipitation and the mean surface temperatures that is in reasonable agreement with observed values from the world-wide GNIP data network (Fig. 7.5). The latitude effect is about

$$\Delta^{18}\delta \approx -0.6\text{‰/degree of latitude} \tag{7.10}$$

for coastal and continental stations in Europe and the USA, and up to -2‰/degree of latitude in the colder Antarctic continent.

The observed relation between monthly temperature and isotopic composition shows much scatter and is not linear, except in the far north. The correlation improves by taking the amount-weighted means. For the North Atlantic and European stations from the GNIP network the relation (t is the surface temperature in °C) is:

$$^{18}\delta = (0.521 \pm 0.014)t - (14.96 \pm 0.21)\text{‰} \tag{7.11}$$

As far as the *palaeoclimatic effect* on the isotopic composition of precipitation and ultimately of groundwater is concerned, the most probable relation between $^{18}\delta$ (or $^{2}\delta$) of precipitation in the region and average temperature at the time of precipitation is similar to the present-day latitudinal temperature dependence at that specific temperature. At 0 °C the effects are approximately 0.8‰/°C and 7‰/°C for $^{18}\delta$ and $^{2}\delta$, respectively.

The theoretical relation between $^{18}\delta$ and $^{2}\delta$, as anticipated in Section 7.1.2.2 and represented by Eq. 7.9, agrees well with the observed values for total precipitation over a large number of years at GNIP stations (Fig. 7.6).

7.1.3.2 *Seasonal effect*

A seasonal temperature pattern is clearly followed by all but some marine stations (Fig. 7.7). The dependence of isotope variations on local temperature (or the closely related parameter

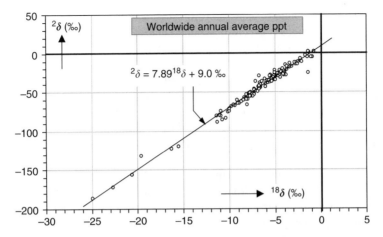

Figure 7.6. Relation between weighted average $^{18}\delta$ and $^{2}\delta$ values for total precipitation over periods of at least one decade from the same stations of the GNIP network as represented in Fig. 7.5. The data reasonably confirm the general validity of the GMWL.

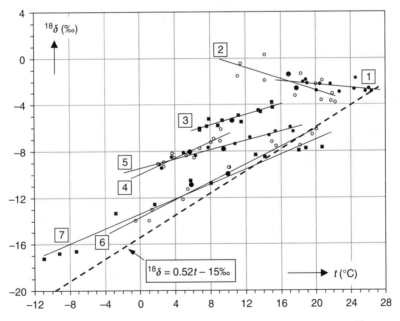

Figure 7.7. Comparison of the latitudinal and the seasonal temperature effect on ^{18}O in precipitation. In both cases the data show the averages over a large number of years. The dashed line refers to the latitudinal effect given by Eq. 7.11, heavy rounds represent the long-term annual averages, and the smaller dots the monthly averages. Numbers indicate the following stations: (1) Midway, Pacific Ocean; (2) Pretoria, South Africa; (3) Valentia, Ireland; (4) Stanley, Falkland Islands; (5) Groningen, The Netherlands; (6) Vienna, Austria; (7) Ottawa, Canada. The opposite seasonal effect for Pretoria may be correlated with high precipitation intensity with often low $^{18}\delta$ values during summer (cf. Section 7.1.4.5).

of precipitable water content) appears as the overriding parameter. The temperature dependence of both δ values is generally smaller than shown by the latitude effect, varying from about 0.5‰/°C for some higher-latitude stations to ultimately 0‰/°C for tropical ocean islands (Fig. 7.7).

From the simple model presented in the preceding section it follows that the seasonal variations of $^2\delta$ and $^{18}\delta$ values for precipitation are bound to be smaller than those of yearly averages with latitude. The reason lies in the $(1/T - 1/T_0)$ factor in Eq. 7.4. During winter, the condensation temperature T is lower than during summer. However, the same is true for T_0, the prevailing average temperatures in the region of evaporation. This partly masks the local temperature effect.

For example, during winter in the Northern hemisphere the highest temperatures are found south of the equator. Water vapour from this region is not easily transported to the Northern hemisphere, however, so the effective evaporation region for the Northern hemisphere has a lower T_0 value than during summer.

Stations situated in a mid-continental setting typically portray a seasonal change in the precipitation isotopic composition. In most cases these variations are correlated with the temperature. For tropical islands, on the other hand, where the vapour source region essentially coincides with the region of precipitation ($T = T_0$) the temperature dependence almost disappears, as is shown in Figs 7.7 and 7.8.

The changes in $^{18}\delta$ are correlated with those of $^2\delta$. However, a local best fit for the $(^{18}\delta, ^2\delta)$ line is in most cases not a meteoric water line in the sense described before, namely one produced by varying degrees of rainout from an air mass of prescribed isotopic character. By and large, most of these effects combine to produce a low-slope $(^{18}\delta, ^2\delta)$ line (i.e. with a slope less than 8) which is not only indicative of the genetic and synoptic history of the rain events, but also reflects the local conditions at the time precipitation occurs. Notably when the slope $\Delta^2\delta/\Delta^{18}\delta$ differs from the value of 8, then a simple linear fit is often not a satisfactory description of the relationships. In this situation a variety of processes is at play, each with its own set of rules concerning the isotope fractionation involved. This is also shown in Fig. 7.10, containing single monthly data for precipitation $^{18}\delta$, contrary to Fig. 7.9 where each point represents an average for one month over a large number of years.

7.1.3.3   *Continental effect*
Precipitation over the ocean, collected at island stations or weatherships, has the characteristics of *a first condensate* of the vapour. The range of most $^{18}\delta$ values is relatively small, between $-0‰$ and $-5‰$ with but little seasonal change in many cases, and a lack of a clear correlation with temperature. The same is true for the $^2\delta$ values (Fig. 7.10).

The *continental effect*, also referred to as the *distance-from-coast effect*, that is, a progressive ^{18}O depletion in precipitation with increasing distance from the ocean, varies considerably from area to area and from season to season, even over a low-relief profile. It is also strongly correlated with the temperature gradient, and depends both on the topography and the climate regime.

During the passage over Europe, from the Irish coast to the Ural mountains, an average depletion of 7‰ in $^{18}\delta$ relative to the Irish precipitation is observed. However, the effect in summer is only about one fourth of the effect in winter. This may be attributed to the re-evaporation of summer rain.

An extreme case of the absence of an inland effect over thousands of kilometres, in spite of strong rainfalls en route, has been reported over the Amazon. This is attributed mainly

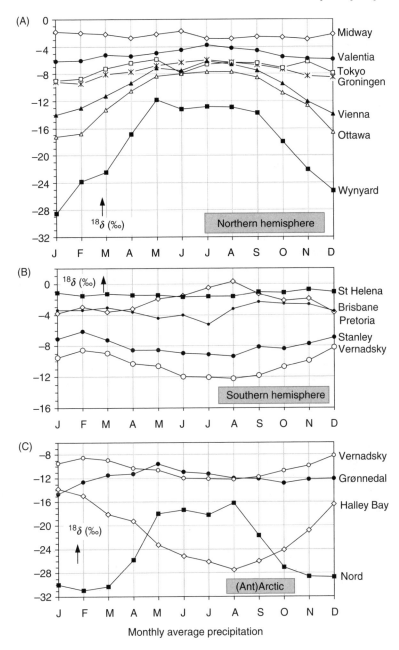

Figure 7.8. Seasonal variations in $^{18}\delta$ shown by weighted averages of monthly precipitation samples (ppt) collected in some typical continental and island or coastal stations from (A) the Northern hemisphere, (B) the Southern hemisphere and (C) some Arctic and Antarctic stations (data from the GNIP network, generally dating from the early 1960s through 1997).

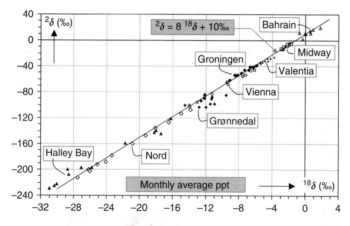

Figure 7.9.   Seasonal influence on the ($^{18}\delta$, $^{2}\delta$) relation for average monthly precipitation at a number of stations; arctic, tropical, coastal and continental (data from the same series as in Fig. 7.8).

to the return flux of the moisture by (non-fractionating) transpiration and thus invalidates the effect of the rainout. However, some of the return flux apparently occurs by evaporation from open waters; this process then results in some change in the isotopic composition.

7.1.3.4   *Altitude effect*

As a rule the isotopic composition of precipitation changes with the altitude of the terrain and becomes increasingly depleted in ^{18}O and ^{2}H at higher elevations. This has enabled one of the most useful applications in isotope hydrology, namely identification of the elevation at which groundwater recharge takes place.

This *altitude effect* is temperature-related, because the condensation is caused by the temperature drop due to the increasing altitude. Due to the decreasing pressure with increasing altitude ($-1.2\%/100$ m), a larger temperature decrease is required to reach the saturated water vapour pressure than for isobaric condensation (Fig. 7.11). Therefore, dN_v/N_v per °C and thus $d\delta/dT$ is smaller than for the isobaric condensation process which produces the latitudinal effect. The molar amount of vapour is proportional to the barometric pressure, b. The relative decrease in vapour content is then given by:

$$dN_v/N_v = dp/p - db/b$$

This decrease is calculated to be $-3.6\%/100$ m (directly caused by a temperature drop of -0.53 °C/100 m) and $+1.2\%/100$ m, respectively. The value of $(dN_v/N_v)dT$ is then $2.4/0.53 = 4.5\%/$°C. From Eqs 7.4 and 7.5 we can then deduce the temperature effect:

$$\frac{d^{18}\delta}{dT} = +0.4\%o/°C \approx -0.2\%o/100 \text{ m} \tag{7.12a}$$

and

$$\frac{d^{2}\delta}{dT} = +3\%o/°C \approx -1.5\%o/100 \text{ m} \tag{7.12b}$$

The most elusive factor is where different air masses with different source characteristics affect the precipitation at the base and crest of a mountain. A prominent case is that of the

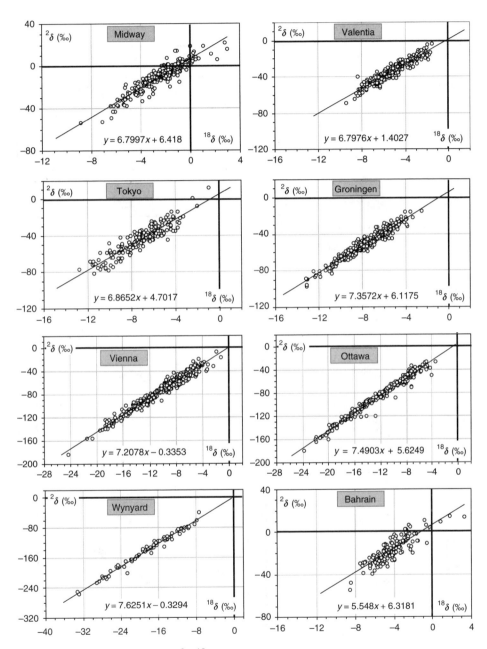

Figure 7.10.   Regression lines of $(^2\delta, ^{18}\delta)$ relations for monthly precipitation samples from marine and continental stations, and stations in tropical, temperate and (Ant)Arctic regions (data from GNIP); d-excess values to be estimated from lines with slope 8 through the data points are variable.

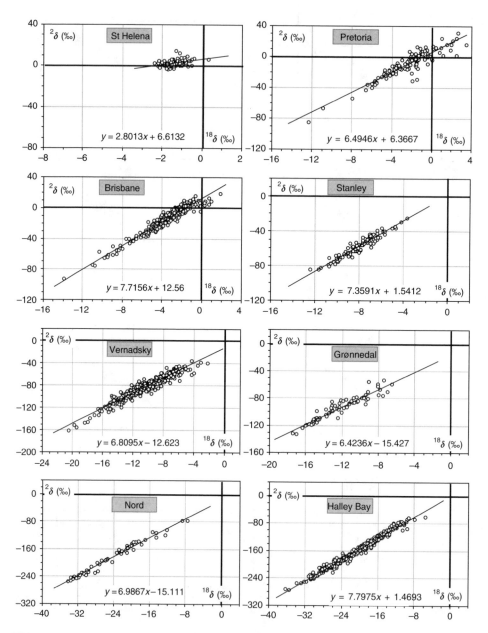

Figure 7.10.   Continued.

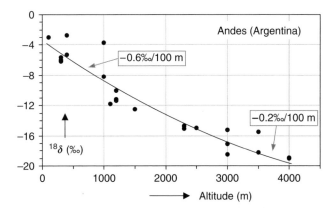

Figure 7.11. Example of the altitude effect on precipitation for the eastern slopes of the Andes mountains, as deduced from samples of shallow groundwater/soil water, collected from springs. The magnitude of the effect increases from −0.2 to −0.6‰/100m.

western slopes of the Andes in South America: precipitation near the crest results predominantly from air from the Atlantic with a long continental trajectory, whereas air from the Pacific Ocean, with predominantly oceanic attributes, affects the precipitation at lower elevations. Under such conditions one encounters apparently anomalously large altitude effects.

An opposite effect results when the long-term samples (e.g. monthly, seasonal or annual composites, such as groundwater or well water) represent different time series at the varying altitudes. As an example can be cited the case of a low-altitude and a high-altitude station in Cyprus. Frontal rains affect both stations and show a normal altitude effect. In the mountain, however, rain also occurs orographically at times when no rain falls in the plain, so that in the composite sample of the mountain station the frontal rains are diluted by precipitation representing a 'first condensate', with relatively enriched isotopic values.

The observed ^{18}O effect generally varies between −0.1‰ and −0.6‰/100 m of altitude, often decreasing with increasing altitude (Fig. 7.11). Values in this range have also been reported for mountain regions in Czechoslovakia, Nicaragua, Greece, Cameroon, Italy and Switzerland. It is obviously a common phenomenon.

In the coastal region of the western United States reported data on the altitudinal effect on $^{2}\delta$ indicated approximate values of −4‰/100 m, whereas values of −2.5‰/100 m were observed in South Western Germany and −1 to −4‰/100 m in Chile.

7.1.3.5 *Amount effect*

A relation between the amount of precipitation and $^{18}\delta$ is often observed. For example, the very strong tropical rainfalls at times of the passage of the Intertropical Convergence Zone (ITCZ), characterised by towering clouds and strong downdrafts, may be extremely depleted in $^{18}\delta$ and $^{2}\delta$, the former by as much as −15‰. Similar, though smaller, effects are observed in thunderstorm-initiated precipitation. In North Western Europe during convective storms, changes in $^{18}\delta$ have been found of −7‰ within 1 hour; examples are shown in Fig. 7.12. The dip in the $^{18}\delta$ curve for seasonal precipitation at Tokyo in June, quite consistent over several years, is probably due to the high rain intensity during this month. The same may be true for the opposite seasonal temperature effect over Pretoria (Fig. 7.8).

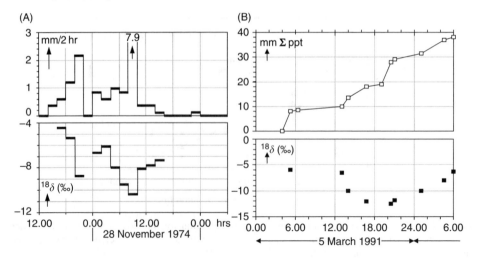

Figure 7.12.   Time sequence of the isotopic composition of precipitation during showers; examples are shown for two cases of convective storms: (A) rain intensity in mm/2 hours; (B) cumulative rain over variable periods. Σ refers to accumulated precipitation.

Among island stations, where temperature variations are small, a dependence of $^{18}\delta$ on rain intensity is observed to the extent of $-1.5‰/100$ mm of monthly precipitation.

On the other hand, small amounts of rain are, as a rule, enriched in the heavy isotopes along typical evaporation lines, especially in more arid regions. This effect obviously results from the evaporation of rain droplets during their fall to the ground. However, no further consistent amount effect is noted for rain intensities in excess of about 20 mm/month.

The conclusion is that one should refrain from generalisations and explore the local amount effect individually for each case, by mounting a special sampling programme.

7.1.3.6   *Interannual variations*
Annual average $^{18}\delta$ values vary from year to year. In temperate climates the values generally do not vary by more than 1‰, and a large part of the spread is caused by variations in the average annual temperature. Figs 7.13 and 7.14 present annual variations for a number of stations; the figures show a spread of the weighted annual precipitation $^{18}\delta$ as well as $^{2}\delta$. Larger variations occur in semi-arid climates, with a less regular rain distribution over time. In those cases only hydrological systems that pool precipitation inputs over many years can be related to the average rain input over extended periods.

7.1.3.7   *Small-scale variations*

7.1.3.7.1   Small-scale spatial variations
It is important to establish for $^{18}O/^{16}O$ tracer studies in which $^{18}\delta$ or $^{2}\delta$ variations in rain are related to those in runoff, that short- and long-term variations of precipitation $^{18}\delta$ with distance have not occurred over the region of interest.

In the temperate climate of North Western Europe it has regularly been observed that the $^{18}\delta$ values of rain samples collected over periods of 8 and 24 hours from three locations within 6 km^2 at equal elevations agree within 0.3‰. Similar results were obtained from the semi-arid climate of Israel.

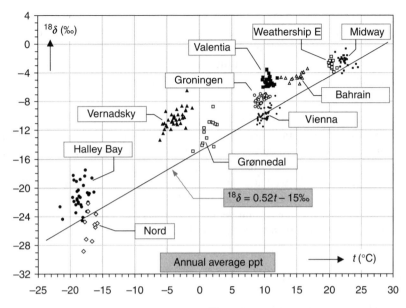

Figure 7.13. Weighted average annual values of $^{18}\delta$ in precipitation (ppt) versus the mean surface air temperature, showing the variations from year to year (data from the GNIP network).

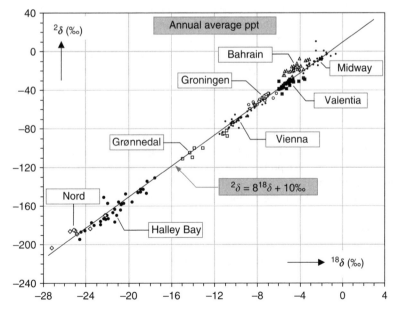

Figure 7.14. Spread of the weighted annual values of the $(^{18}\delta, {}^{2}\delta)$ relation for precipitation (ppt) (data from the GNIP network).

At larger distances, especially with single convective storms, larger differences are to be expected because of the increasing depletion of ^{18}O and ^{2}H. On the other hand, over periods of months $^{18}\delta$ in rain may show similar patterns, even over distances of a few hundred kilometres. Despite the similarities, single monthly samples might differ significantly, although the average values are only slightly ($\approx 1\%_o$) different. The $^2\delta$ and $^{18}\delta$ values probably closely follow the MWL.

7.1.3.7.2   Small-scale temporal variations

Rapid variations in $^{18}\delta$ of rain can occur. In Section 7.1.4.5 we have mentioned such variations in samples collected during successive two-hour periods. The magnitude of this effect depends on the character of the precipitation process, such as convective storms or weather fronts. Fig. 7.12 showed an example, in the framework of the amount effect. The relatively large isotopic variations in rain offer the possibility to correlate precipitation, runoff and groundwater within certain periods and in certain regions (Chapter 8; see also Volume III on Surface Waters in the UNESCO/IAEA series).

7.2 ^{3}H IN PRECIPITATION

7.2.1   Characteristics and distribution of ^{3}H

The radioactive isotope of hydrogen of mass 3 (^{3}H, or *tritium*) has a half-life of 12.32 years, and thus a lifetime comparable to many hydrological processes. Natural ^{3}H is introduced to the hydrological cycle mainly in the atmosphere, where it is produced by the bombardment of nitrogen nuclei by cosmic-ray neutrons.

From the early 1950s anthropogenic sources, especially from nuclear tests in the atmosphere, overshadowed the natural production for more than a decade. More recently, large amounts of tritium are being produced in the cooling-water systems of nuclear power stations, and subsequently released into adjacent surface waters (predominantly rivers).

Fig. 7.15 shows the ^{3}H concentration in monthly precipitation at Vienna (data from the GNIP network, see Section 7.1). Data from the Ottawa station indicate that the ^{3}H increase started already during the 1950s.

From known-age wine samples it has been estimated that the natural ^{3}H content of rain before the nuclear test series in the 1950s began was about 5 TU in central Europe. At present the ^{3}H content of precipitation is again approaching this natural value. The production of ^{3}H takes place preferentially in the upper troposphere and lower stratosphere. It is introduced into the hydrological cycle following oxidation to tritiated waters (^{3}H^{1}HO), seasonally leaking down into the troposphere mainly through the tropopause discontinuity at mid-latitudes.

At the peak concentration of ^{3}H during spring 1963, the ^{3}H content of precipitation in the Northern hemisphere was about 5000 TU. The pattern shown in Fig. 7.15 is repeated at most Northern hemisphere stations, albeit with slightly varying amplitudes and phase shifts. The notable feature of this curve is a yearly cycle of maximum concentrations in spring and summer and a winter minimum, with typical concentration ratios of 2.5–6 between maximum and minimum values. The annual cycle is superimposed on the long-term changes which have ranged over three orders of magnitude since 1952.

There is a marked latitude dependence: concentrations are highest north of the 30th parallel, with values lower by a factor of 5 or so at low-latitude and tropical stations.

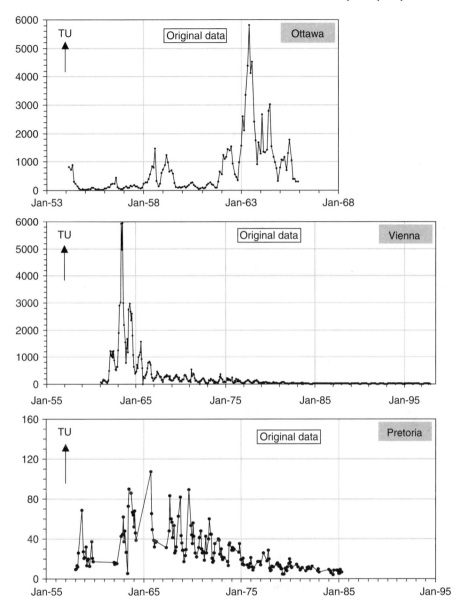

Figure 7.15. ^{3}H in monthly precipitation samples of stations representative for the Northern (Ottawa, Vienna) and the Southern hemispheres (Pretoria) (data from the GNIP network, valid for the year of sample collection). 'Original data' means that the data have not been corrected for radioactive decay, and therefore refer to the date of sample collection.

In the Southern hemisphere (represented by Pretoria) the yearly cycle is displaced with the season by half a year, and the mean ^{3}H levels in atmospheric waters are lower than at comparable north-latitude stations. This is a reflection of the predominant northern location of weapon testing sites and the slow inter-hemispheric transport of tracers.

Consequently, in the Southern hemisphere the increase has been a factor of 10 to 100 smaller (Fig. 7.15), because of the equatorial barrier in the global air mass circulation and the fact that the annual ^{3}H injected during spring from the stratosphere into the troposphere is removed from the latter very efficiently within weeks.

7.2.2   Hydrological aspects

It is of interest to be able to estimate ^{3}H concentrations in the underground, in order to gain a first impression of the hydrological situation, and of the necessity and specific conditions for sample collection. Since the ^{3}H content of precipitation has decreased to almost its natural level, part of the applications exploited during the 1960s and 1970s are no longer possible. Nevertheless, excess ^{3}H is still present in the ground and in surface waters. Therefore, knowledge of possible variations is still relevant.

In the next sections we present a broad summary of ^{3}H variations in precipitation.

7.2.2.1   *Long-term recovery of natural ^{3}H levels*
There are two reasons why ^{3}H in precipitation has returned quickly to its natural level, at least in comparison with ^{14}C in atmospheric CO_2:

1. ^{3}H has a relatively *short half-life* of 12.32 years. This means that by decay alone the peak concentration from 1963 would become reduced by a factor of 2^{10} ($\approx$ 1000 in a period of about 120 years (=10 half-lives) provided no further atmospheric nuclear tests were made. In the period of three half-lives since 1963, radioactive decay has reduced the peak concentration by a factor of 8. Fig. 7.16 shows the data for Vienna and Pretoria from Fig. 7.15 when corrected for radioactive decay.
2. The *precipitation process* is the main mechanism for removing ^{3}H from the atmosphere over the continent. The atmospheric water circulation through oceanair exchange is very vigorous, with the water content of the troposphere being replaced about every 10 days. Therefore, bomb ^{3}H is rapidly transported to the ocean, while ^{14}C is not (Chapter 6). On the other hand, most ^{3}H produced by fusion bombs was injected into the stratosphere, from which it has only gradually been leaking back to the troposphere, thus extending the presence of ^{3}H in the atmosphere.

Because of the short turn-over time for atmospheric water and the fact that not much ^{3}H is left in the stratosphere, the tropospheric ^{3}H concentration has decreased to the relatively stable original level. It is only since the 1970s, following a long period of limited nuclear test activity, that the Northern and Southern hemisphere ^{3}H concentrations are becoming comparable.

7.2.2.2   *Seasonal variations in ^{3}H*
The seasonality of the stratosphere-to-troposphere transport results in the marked seasonal cycle in the ^{3}H content of precipitation (Fig. 7.16), opposite in phase between the Northern and Southern hemispheres.

Fig. 7.17 contains two graphs of separate 'winter' and 'summer' ^{3}H patterns, one containing the weighted means for the months October–March, taken as representative for winter, and April–September for the summer months in the Northern hemisphere, whereas similar data for Melbourne represent the Southern hemisphere. This phenomenon is caused by temporary mixing between the stratosphere and troposphere at high latitudes in early

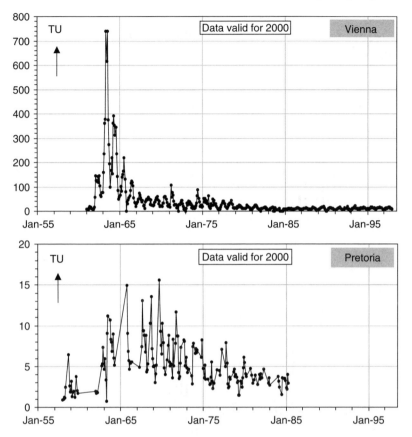

Figure 7.16. ^{3}H in precipitation at Vienna (representing the Northern hemisphere) and Pretoria (representing the Southern hemisphere), corrected for radioactive decay during the period between the moment of sampling and the year 2000 (original data from Fig. 7.15); in other words, the data in the graphs are valid for the year 2000.

spring so that ^{3}H, originally injected into the stratosphere by nuclear explosions, can return to the troposphere.

For hydrologists this is a very important aspect of the temporarily elevated ^{3}H levels from about 30 years ago: *the infiltration of precipitation appears to be not distributed evenly over the year*. Groundwater recharge generally occurs after heavy rainfall and without significant (evapo)transpiration by the vegetation. In moderate climates infiltration is therefore limited to the 'winter-period' with relatively low ^{3}H values. This point is further discussed in Chapter 8.

7.2.2.3   *Geographical variations in ^{3}H*
In Figs 7.15 and 7.17 we have compared the ^{3}H content of precipitation at different stations. As we have seen, the most striking point is that the very high ^{3}H levels are confined to the Northern hemisphere: in the Southern hemisphere, the ^{3}H content increased by a factor of hardly more than 10 above the pre-bomb levels. The variations at stations in the Northern hemisphere show generally the same pattern. However, the ^{3}H concentrations themselves may be quite different from one station to another.

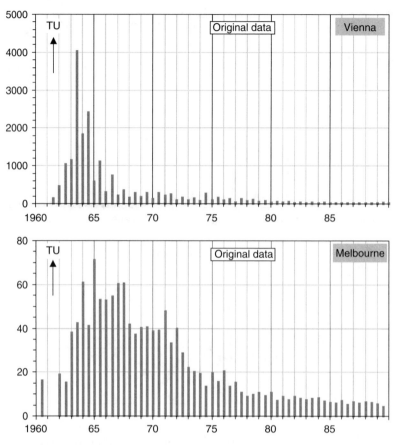

Figure 7.17. ^{3}H values for 'winter' (October–March) and 'summer' (April–September, between scale divisions) precipitation in Vienna, representing the Northern hemisphere and similar values for Melbourne at the Southern hemisphere. The seasonal cycle for the latter series is less pronounced (weighted averages from the GNIP network). In the Southern hemisphere the ^{3}H (and ^{14}C) maxima have a phase shift of half a year, because the leak between stratosphere and troposphere occurs in early spring. Fig. 9.2 shows an example of the absence of high 'summer' values in the unsaturated zone in a dune area in North Western Europe, meaning that here the 'summer' rain did not reach the groundwater table (see also Volume IV of the UNESCO/IAEA series on Groundwater; Section 5.1.4.2) ('original data' means: data valid for the year of sample collection).

By an effect similar to the continental effect and the (small) seasonal effect for ^{18}O and ^{2}H, low ^{3}H values are found near the ocean. A large fraction of the local water vapour, and thus of the precipitation, consists of oceanic vapour which is low in ^{3}H.

7.2.2.4 *Small-scale ^{3}H variations*

Local variations are likely to be small, because the ^{3}H content in rain is not influenced by temperature variations (as are ^{18}O and ^{2}H). Although ^{3}H is also fractionated during evaporation and condensation processes – at twice the extent as ^{2}H – the variations involved are in the order of 16% (twice 80‰), equivalent to just one year of radioactive decay. Therefore, they cannot be clearly distinguished in hydrological reality and are thus neglected.

Under normal conditions we would not expect significant variations in the ^{3}H content of the vapour within one air mass.

From a series of ^{3}H data for monthly precipitation from 15 stations within the small country of the Netherlands (about 30 000 km^2), it is concluded that between the stations there are no significant regular differences in seasonal effect; existing differences over this region are irregular and small. Averaged over the year, however, a small continental effect is apparent.

Another example over a longer period is presented by four stations within 50 km around Vienna. The seasonal variations are in phase and the yearly averages are comparable. Discrepancies between the stations do not appear to be systematic.

Rapid fluctuations in ^{3}H can be expected, if in a short period different air masses contribute to the precipitation at a site. Precipitation collected during a severe convective storm showed no significant differences within a period of 30 minutes, although significant variations in ^{18}O and ^{2}H were noted. A quick change in ^{3}H might occur during the passage of another air mass, for example, correlated with a cold front. Large differences in ^{3}H content have also been observed within a hurricane, resulting from complex meteorological conditions.

CHAPTER 8

Isotopes in surface water

Basically, there are two different types of application of natural isotopes in surface water hydrology:

1. The natural variations in ^{18}O, ^{2}H and ^{3}H can be considered as *tracers* and can as other tracers be used in surface-water applications. The interaction of surface- and groundwater plays an important role in this application.
2. At the surface, such physical and geochemical processes as evaporation and exchange may alter the isotopic composition of the water and of its dissolved carbon. For studying these processes, ^{18}O and ^{2}H, and ^{13}C are especially suitable.

A general introduction to hydrological aspects of the Earth's freshwater masses is given in Volume III of the UNESCO/IAEA series. In this chapter we restrict ourselves to the isotopic abundances, particularly concerning ^{18}O, ^{2}H, ^{3}H, ^{13}C and ^{14}C.

8.1 ISOTOPE FRACTIONATION DURING EVAPORATION

8.1.1 Effective fractionation of ^{18}O and ^{2}H

In the further discussions in this section we discuss and calculate isotopic changes in surface water during evaporation. Because a net evaporation requires the atmospheric vapour to be undersaturated, these changes are not governed by *equilibrium* fractionation factors. Therefore, we will apply the *real* or *effective* fractionation factors α for ^{18}O and similarly for ^{2}H. In order to be more specific, we have to consider the exchange between the water and the atmosphere, rather than merely the removal of water vapour from the liquid. Fig. 8.1 clarifies this condition.

The rare isotope mass balance gives:

$$RN + \alpha \downarrow R_v[h/(1-h)]dN = (R - dR)(N - dN) + \alpha \uparrow R[1/(1-h)]dN$$

where R and R_v are the isotopic composition of the water and the atmospheric vapour, respectively, h is the relative humidity, and $\alpha \uparrow$ and $\alpha \downarrow$ are the one-way (kinetic) fractionation factors of the 'evaporation' and the 'condensation' process. According to Eq. 2.17:

$$\alpha \uparrow /\alpha \downarrow = \alpha_{v/l} \tag{8.1}$$

as the equilibrium fractionation of the vapour relative to the water.

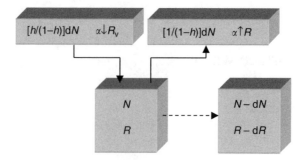

Figure 8.1.   Model representing a well-mixed reservoir subjected to a net evaporation and exchange with the ambient water vapour (cf. Fig. 3.6).

The resulting differential equation is:

$$\frac{dN}{N} = \frac{(1-h)dR}{(\alpha\uparrow -1+h)R - h\alpha\downarrow R_v} \tag{8.2}$$

for which the solution is:

$$R = \Phi R_v + [R_{10} - \Phi]\left(\frac{N}{N_0}\right)^{[\alpha\downarrow\alpha/(1-h)]-1} \tag{8.3}$$

where

$$\Phi = \frac{\alpha\downarrow h}{\alpha\downarrow\alpha - 1 + h}$$

Experiments have shown that

$$1/^2\alpha\downarrow = 1.010 \tag{8.4a}$$

$$\text{and}\quad 1/^{18}\alpha\downarrow = 1.008 \tag{8.4b}$$

The isotopic composition of the *net amount* of water at any time leaving the water body is:

$$R_{net} = \frac{\alpha\uparrow R\frac{1}{1-h}dN - \alpha\downarrow R_v\frac{h}{1-h}dN}{dN} = \alpha\downarrow(\alpha_{v/l}R - hR_v)/(1-h)$$

so that the *effective fractionation* is:

$$\alpha_{eff} = \frac{R_{net}}{R} = \frac{\alpha\downarrow}{1-h}(\alpha - h(1+\delta_v - \delta)) \tag{8.5}$$

If water evaporates in contact with the atmosphere, ^{18}O is, compared to ^2H, more enriched than that according to the factor of 8 in the relation for the MWL. This process is similar to the equatorial evaporation of the oceans, which was discussed in Section 7.1.1. The reason for the anomalous $^{18}\delta - ^2\delta$ relation is to be found in the fact that the relative humidity is below 100%. A quantitative insight is presented by applying the Rayleigh evaporation model (Section 3.2). Nature provides a few essentially different cases, depending on the presence of an inflow to and/or an outflow from the evaporating water mass.

8.1.2 Relation between isotopic and chemical composition

The consequences of evaporation for the isotopic composition of water have been calculated in Section 2.11, resulting in:

$$\delta = \left(\frac{N}{N_0}\right)^{\alpha-1} - 1 \qquad \text{(cf. Eq. 3.8)}$$

where the original δ value (δ_0) is taken $= 0$. Contrary to the decrease in the amount of water during evaporation is the increase of chemical compounds dissolved in the water, such as salt. Therefore, the relation between the salinity (salt content) S and the isotopic composition is:

$$\delta = \left(\frac{S}{S_0}\right)^{1-\alpha} - 1 \tag{8.6}$$

N/N_0 is decreasing, while S/S_0 increasing during evaporation: $(N/N_0) = (S_0/S)$. Consequently, the exponent has to change from $(\alpha - 1)$ to $(1 - \alpha)$. The result is shown in Fig. 8.2 for an ε value of $-10‰$.

8.1.3 Relation between ^{18}O and ^2H by evaporation

From Eq. 8.2 we can calculate the ratio between the $^2\delta$ and $^{18}\delta$ changes during evaporation:

$$\frac{dR/dN \text{ for } ^2H}{dR/dN \text{ for } ^{18}O} = \frac{[(\alpha \downarrow \alpha_{v/l} - 1 + h)(1 + \delta) - \alpha \downarrow h(1 + \delta_v)] \text{ for } ^2H}{[(\alpha \downarrow \alpha_{v/l} - 1 + h)(1 + \delta) - \alpha \downarrow h(1 + \delta_v)] \text{ for } ^{18}O} \tag{8.7}$$

Using Eq. 8.7 with $^2\alpha_{v/l}$ and $^{18}\alpha_{v/l} = 0.9311$ and 0.99111, respectively, at $25\,°C$, $h = 0.6$, and realistic values for the δ values under tropical conditions, the resulting ratio is in the order of:

$$\frac{\Delta^2\delta}{\Delta^{18}\delta} = 4.5 \pm 0.5 \tag{8.8}$$

The process is schematically illustrated in Fig. 8.3.

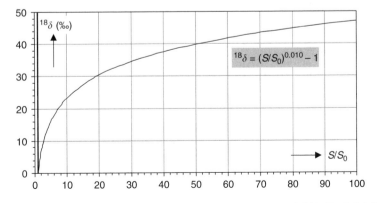

Figure 8.2.   Relation between salinity and isotopic composition for $\alpha = 0.990$ (Eq. 8.6) (cf. Fig. 3.4).

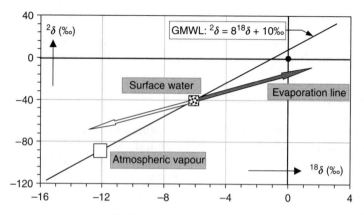

Figure 8.3.   Relation between the $(^{18}\delta, {}^{2}\delta)$ values of meteoric water which undergoes evaporation, the vapour leaving the water and the residual water following evaporation, described by the *evaporation line*, compared with the relationship between atmospheric water and precipitation described by the meteoric water line (cf. Fig. 5.18). The relatively 'light' (depleted) water vapour leaves the water reservoir (open arrow), causing the residual water to become enriched (grey arrow).

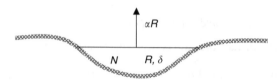

Figure 8.4.   Evaporation from an isolated lake with oxygen isotope fractionation α. The resulting isotopic change is shown in Fig. 3.4.

8.1.4   Evaporation processes in nature

Some Rayleigh processes we dealt with in this section have been treated in detail in Chapter 2. From these only results will be mentioned.

8.1.4.1   *Evaporation of an isolated water body*

This process, outlined schematically in Fig. 8.4, has been treated mathematically in Section 2.11.1 resulting in:

$$\frac{R}{R_0} = \left(\frac{N}{N_0}\right)^{\alpha-1} \qquad \text{(cf. Eq. 3.7)} \tag{8.9}$$

or in δ values with respect to the standardised reference:

$$\delta = (1 + \delta_0)(N/N_0)^{\varepsilon} - 1 \qquad \text{(cf. Eq. 3.8)} \tag{8.10}$$

Fig. 3.4 shows the ^{18}O enrichment of the lake water, if the fractionation is -10%.

8.1.4.2   *Evaporation of a well-mixed water body with constant inflow and no outflow*

Several cases are known of lakes where the evaporation keeps pace with the river input (Dead Sea, Lake Chad). We will therefore limit our discussion to the steady-state in which

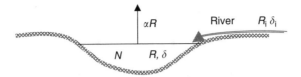

Figure 8.5.   Evaporation from a well-mixed lake with a constant river inflow.

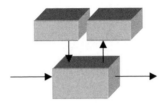

Figure 8.6.   Model representation of a well-mixed reservoir with river input and output and evaporation (see Fig. 3.8 and the resulting enrichment shown by Fig. 3.9).

the isotopic composition of the lake water is constant. A condition for our present treatment is that river input is constant, and that the lake water is well mixed and of homogeneous isotopic composition (Fig. 8.5) (cf. Section 3.2.3). The final situation, in which the water inflow is balanced by the evaporation, is simply understood by realising that the steady-state requires the isotopic composition of the vapour leaving the water to be equal to that for the river inflow:

$$\alpha R_{\text{steady state}} = R_i \quad \text{or} \quad R_{\text{steady state}} = R_i/\alpha \tag{8.11}$$

so that

$$\delta_{\text{steady state}} \approx \delta_i - \varepsilon \tag{8.12}$$

where $^{18}\varepsilon = -9$ to $-10\%\!o$ and $^2\varepsilon = -70$ to $-80\%\!o$, depending on temperature (Table 5.4). The resulting $\delta > \delta_i$ because $\varepsilon < 0$.

8.1.4.3   *Evaporation of a well-mixed water body with constant inflow and outflow*
This process has been discussed in Section 3.2.5 and is clarified by Fig. 8.6.

The enrichment depending on the water loss by evaporation is shown in Fig. 3.9. The $(^{18}\delta, {}^2\delta)$ relation is characterised by a slope 4 to 5, also depending on conditions of temperature and relative humidity.

8.1.4.4   *Evaporation of an isotopically inhomogeneous lake with*
        a river inflow and with or without discharge
The preceding paragraph unrealistically suggests that when lake evaporation is balanced by river input, the lake water is expected to have an isotopic composition which is the fractionation apart from that of the river water. For example, $^{18}\delta_{\text{steady state}} = {}^{18}\delta_{\text{river}} - {}^{18}\varepsilon_{\text{eff}} = -4-(-9\%\!o) = +5\%\!o$. However, lakes are not well-mixed water bodies. The actual enrichments observed result from a progressive enrichment in the lake, discontinuously illustrated by Fig. 8.7; values can be reached of up to $+30\%\!o$, as in Lake Chad and in the Dead Sea.

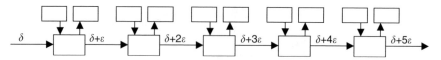

Figure 8.7.   Schematic representation of stepwise progressive enrichment of river- or lake water.

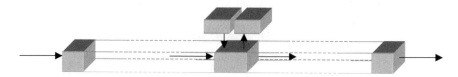

Figure 8.8.   Schematic representation of the model shown by Fig. 3.8 for a continuous progressive enrichment of river- or lake water.

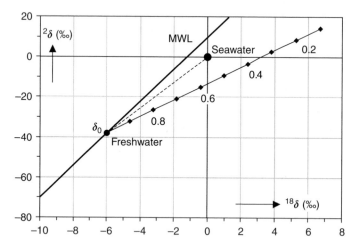

Figure 8.9.   Model for calculating the shift in $^{18}\delta$ and $^{2}\delta$ of river water subjected to evaporation during a relatively long residence time in its estuary (see UNESCO/IAEA Volume I, section 4.4.5). The relevant parameters are similar to the example of Fig. 3.9. MWL denotes the Global Meteoric Water Line. The values along the 'evaporation line' refer to the fractions of water remaining after evaporation. The dashed line represents estuarine mixing.

The process of a continuous enrichment is outlined schematically in Fig. 8.8 and treated mathematically in Section 3.2.5.

In terms of the evaporated fraction, the result of the model described by Fig. 3.8 is shown by Fig. 3.9. The combined ^{18}O and ^{2}H enrichment depends on the values of the various parameters; a realistic outcome is shown in Fig. 8.9.

Fig. 8.10 summarises the types of possible variation in the $(^{18}\delta, ^{2}\delta)$ relation. The meteoric water lines are shown to depend on the relative humidity in the source region of the atmospheric vapour (cf. Eq. 8.5). For example, in the Eastern Mediterranean the Regional Meteoric Water Line observed is:

$$^{2}\delta = 8\,^{18}\delta + 20‰ \qquad\qquad (8.13)$$

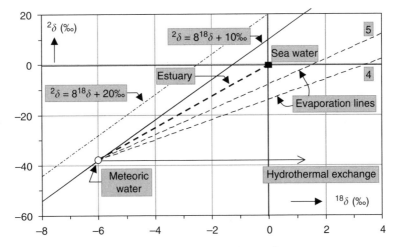

Figure 8.10.   Diagram illustrating the relations between ^{18}O and ^2H variations in natural waters: the Global Meteoric Water Line, the mixing line for fresh- and sea water in estuaries, the evaporation lines with slopes 4 and 5 (Eq. 8.8) starting from an arbitrary meteoric water, and the oxygen isotopic exchange in hydrothermal water at high temperatures.

In evaporating surface water, $^{18}\delta$ and $^2\delta$ are shifted to higher values along evaporation lines with slopes of 4 to 5 depending on the relative humidity and the degree of evaporation.

Underground water-rock interaction at elevated temperatures can cause an isotopic shift in $^{18}\delta$ while $^2\delta$ remains unaffected. Examples of this phenomenon have been reported for thermal waters in Italy and New Zealand.

8.2   OBSERVED ^{18}O AND ^2H VARIATIONS IN SURFACE WATER

8.2.1 ^{18}O and ^2H in large rivers

It is evident that the isotopic composition of river water is related to that of precipitation. In small drainage systems, $^{18}\delta$ for the runoff water is equal to that for the local or regional precipitation. In large rivers, transporting water over large distances, there may or may not be a significant difference in $^{18}\delta$ between the water and the average precipitation at the sampling location. Examples are given in Fig. 8.11.

Apart from a possible continent effect, $^{18}\delta$ for the River Rhine is influenced by the altitude effect in precipitation: a significant part of the Rhine water is melt water from the Swiss Alps ($^{18}\delta \approx -13‰$) (Fig. 8.13). For the same reason, a lack of agreement between $^{18}\delta$ in runoff and regional precipitation is observed for the rivers Caroni (Venezuela), Jamuna (Bangladesh) and Indus (Pakistan). The sharp decrease in $^{18}\delta$ for the Caroni during April 1981 is due to a large discharge of precipitation over the Andes mountains. On the other hand, the rivers Meuse (Netherlands), Mackenzie (Canada) and Parana (Argentina), for example, represent average precipitation $^{18}\delta$ values.

Seasonal variations in $^{18}\delta$ are obvious in both Figs 8.11 and 8.13, in addition to differences in average $^{18}\delta$ level. In large rivers, $^{18}\delta$ increases during summer are not caused by evaporation. For example, the ($^{18}\delta$, $^2\delta$) relation for the rivers Rhine and Meuse follow

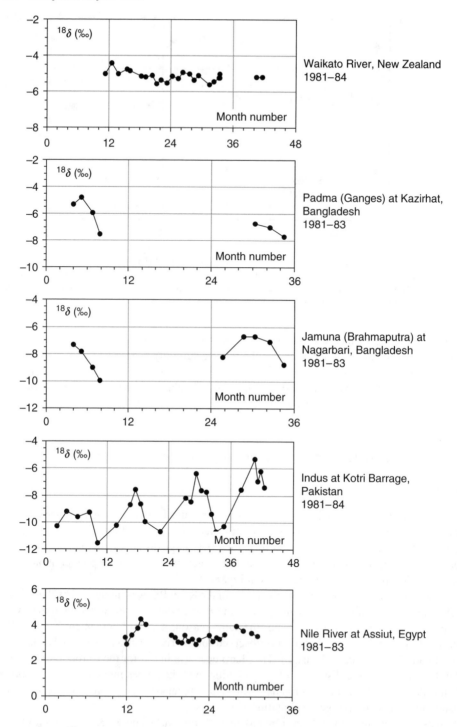

Figure 8.11. $^{18}\delta$ for some major rivers in the world. Part of the data was obtained in co-operation with participants in the SCOPE/UNEP project on *Transport of Carbon and Minerals in Major World Rivers*.

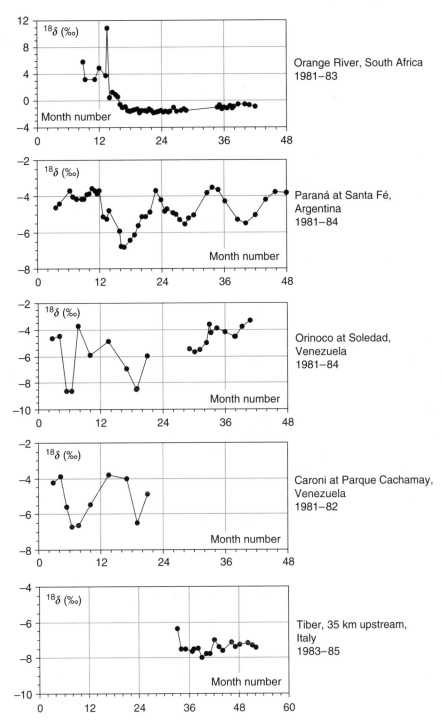

Figure 8.11.   Continued.

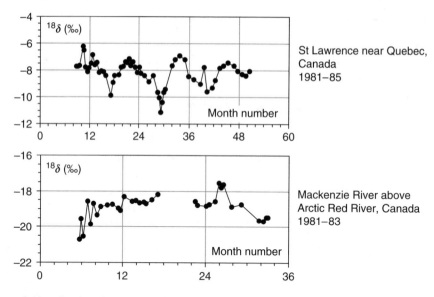

Figure 8.11. Continued.

the meteoric water line (Eqs 5.36, 7.9 and 7.11) and do not show the lower slope typical of evaporated waters (Fig. 5.38 and Section 8.1.2). The origin of the seasonal variation is therefore attributed to the seasonal variations in precipitation. Large rivers thus contain not only old groundwater with a constant isotopic composition, but also a relatively fast component which can be ascribed to surface runoff.

The inverse seasonal effect in the Dutch rivers Rhine (and IJssel) and in the rivers Caroni, Padma, Jamuna and Indus, is caused by a relatively large *melt water* component during spring and summer. If for the Rhine we assume that without this component the water would show a seasonal variation similar to that of the other Dutch rivers, and that $^{18}\delta$ for the winter discharge from South Western Germany is $-9‰$, the melt water contribution to the Rhine is about 30% of its total discharge during summer. Here we use the simple (approximated) additivity of the δ values for the constituting water components:

$$^{18}\delta_{mixture} = f_1^{18}\delta_{comp.1} + f_2^{18}\delta_{comp.2} \qquad \text{(cf. Eq. 3.3)} \qquad (8.14)$$

As discussed in Section 7.1.2.1, $^{18}\delta$ and $^{2}\delta$ variations in meteoric water are generally closely related through the equation of the meteoric water line: $^{2}\delta = 8^{18}\delta + 10‰$. Therefore, our discussion on ^{18}O variations in surface water also applies to ^{2}H variations. The ($^{18}\delta$, $^{2}\delta$) relation in Fig. 8.12 shows that most rivers contain meteoric water ($^{18}\delta$ in Fig. 8.11). One river, however, the Orange River in South Africa, evaporates significantly.

A useful application of differences in average $^{18}\delta$ for river water and local groundwater is determination of the extent of river-bed infiltration. This refers especially to regions where drinking-water supply stations are situated near polluted rivers (e.g. Rhine).

Another application of $^{18}\delta$ differences in river waters is estimating the relative contribution of a tributary to the main river. Fig. 8.13 shows that the river IJssel, a branch of the Rhine, contains an additional component of regional discharge ($^{18}\delta = -7.5‰$).

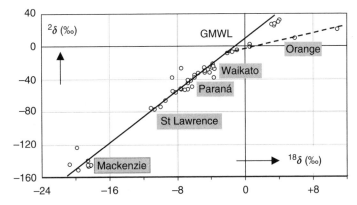

Figure 8.12.   Relation between $^{18}\delta$ and $^{2}\delta$ for a choice of samples from a few rivers. For the rivers in temperate regions evaporation is insignificant, contrary to the Orange River in South Africa.

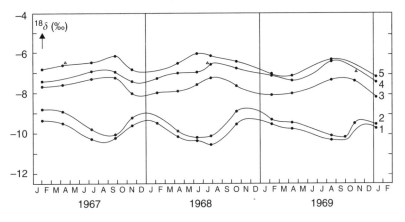

Figure 8.13.   Some *rain water* rivers in the Netherlands (3 = Meuse, 4 = Vecht, 5 = Zwartewater) show a seasonal variation opposite to the *melt water* rivers (1 = Rhine, 2 = IJssel, a branch of the Rhine, originating in part from the Swiss Alps).

From the difference between both rivers, the local component is deduced to vary between 10 and 30%.

8.2.2 ^{18}O and ^{2}H in small rivers and streams

Small rivers respond faster to changes in precipitation $^{18}\delta$ than do large rivers, because the surface runoff component of the discharge is more direct and pronounced. This is evident from the decrease in concentration of various chemical compounds (e.g. in the alkalinity) during periods of high rainfall. The isotope effect is qualitatively shown in Fig. 8.14 for a small river (Drentse Aa, the Netherlands).

Fig. 8.15 illustrates another example of the effect of $^{18}\delta$ variations in rain on those in the runoff from a small stream. Preliminary investigations showed that a sampling frequency of 8 hours was not sufficient in this basin (catchment about 6 km^2).

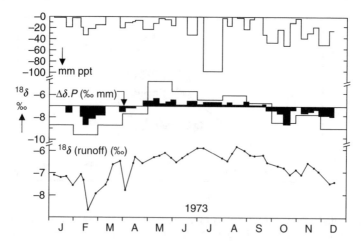

Figure 8.14.   Qualitative comparison of runoff and precipitation from $^{18}\delta$ variations in a small river (Drentse Aa, the Netherlands), by correlation with the precipitated 'amount of ^{18}O tracer', (black area) calculated by Eq. 8.15 from the precipitation intensity (in mm ppt per month) and $^{18}\delta$ for the monthly rain.

In a second period (Fig. 8.15) use was made of automatic samplers for the rain (2 hours-samples) as well as the runoff (once per 2 hours). In this way the runoff hydrograph could be analysed for the base-flow and some fast components; that is, *hydrograph separation* in surface runoff, subsurface runoff and bank storage. Furthermore, it was possible to determine the percentages of infiltration and direct runoff caused by the rainfall, by measuring and comparing the peak surfaces of the $\Delta\delta \times Q$ and $\Delta\delta \times P$ curves.

According to Section 2.9, the concentration of the ^{18}O tracer is defined as:

$$[H_2{}^{18}O] - [H_2{}^{18}O_{base}] = {}^{18}R_r\{{}^{18}\delta - {}^{18}\delta_{base}\} \quad \text{(cf. Eq. 2.48)} \qquad (8.15)$$

$^{18}R_r$ is the isotope ratio for the international reference (VSMOW). To compare the amount of runoff with that of precipitation, we have introduced the *amount of tracer*, defined as:

$$^{18}R_r\{{}^{18}\delta - {}^{18}\delta_{base}\} \times P \text{ (or } Q) = {}^{18}R_r\Delta{}^{18}\delta \times P \text{ (or } Q) \quad \text{(cf. Eq. 2.49)} \qquad (8.16)$$

where P denotes the amount of precipitation, Q the amount of runoff.

Comparing these leads to calculating the ratio:

$$\frac{\sum_i(\Delta{}^{18}\delta_Q Q)_i}{\sum_j(\Delta{}^{18}\delta_P P)_j} \quad \text{(cf. Eq. 2.50)} \qquad (8.17)$$

for a number of time periods i and j (i may equal j, depending on the question). In these examples the percentages of direct runoff compared to the precipitation during that period are about 13 and 25%, respectively.

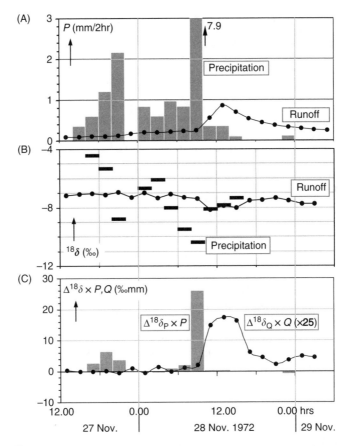

Figure 8.15. ^{18}O variations for precipitation and total runoff (B) in a small catchment area of 6 km^2, in the Eastern Netherlands, related to precipitation (P in mm) and runoff quantities (Q in mm over the catchment area) (A). (C) presents the 'amount of ^{18}O' in the precipitation and runoff calculated from $(\delta_P - \delta_0) \times P = \Delta\delta_P \times P$ and $(\delta_t - \delta_0) \times Q = \Delta\delta_Q \times Q$. The base-flow value ($\delta_0$) for ^{18}O was chosen to be $-7.1‰$. For the sake of clarity the runoff curve in figure (C) is increased by a factor of 25.

8.2.3 ^{18}O and ^2H in the sea

Based on the first measurements of sea water, the primary reference sample, SMOW, was defined in Section 5.2.3. Values of $^{18}\delta$ for the various water masses are listed in Table 8.1.

Larger variations and deviations from $^{18}\delta = 0‰$ are found where:

- Sea water is subject to evaporation (Persian Gulf: $+2‰$; Mediterranean Sea: $+1‰$; equatorial surface waters: $+0.7 \pm 0.2‰$).
- Coastal waters contain variable amounts of freshwater runoff from continental rivers.
- Polar waters contain variable amounts of melt water and freshwater runoff primarily from Siberia.

In the latter two cases $^{18}\delta$ is related to the salinity. The ^{18}O enrichment in an evaporating water body and the correlation with salinity are discussed in Sections 8.1.1 and 8.1.2.

Table 8.1.   Values for the hydrogen and oxygen isotopic composition and salinity of deep water in various oceans and marine basins (chapter 3 in Volume III of the UNESCO/IAEA series).

Ocean	$^{18}\delta$ (‰)	$^{2}\delta$ (‰)	S (‰)
Arctic		$+2.2 \pm 1.0$	
North Atlantic	$+0.12$	$+1.2 \pm 0.8$	34.9
South Atlantic		-1.3 ± 0.6	
Pacific	-0.2 ± 0.3	1.4 ± 0.4	34.7
Antarctic	-0.2 to -0.45	-0.9 to -1.7	34.7
Indian	-0.2 ± 0.2		34.7 ± 0.2
Mediterranean	$+1.5 \pm 0.2$		39.0 ± 0.2
Baltic Sea	$> -7 \pm 1^*$		$>4^*$
Black Sea	-3.3 ± 0.2		
Red Sea	$+2.5$		

*Values for the freshwater component; higher values are found with admixture of sea water.

During past glacial periods a large amount of water was withdrawn from the oceans and deposited as vast ice sheets in the polar regions and on the North American and European continents. This ice is expected to have had a low ^{18}O content. Consequently, $^{18}\delta$ for the glacial oceans must have been higher. Depending on the volume of polar ice and its $^{18}\delta$ value ($-15‰$), the estimated changes in $^{18}\delta$ during the last glacial/interglacial transition at about 12.000 to 9.000 BP are between -0.5 and $-1.0‰$. The glacial oceans therefore had a higher $^{18}\delta$ value of $+0.5$ to $+1.0‰$, the main reason for the $^{18}\delta$ amplitude of deep-sea palaeotemperature records (Fig. 5.13).

8.2.4 ^{18}O and ^{2}H in estuaries and coastal waters

In a region of mixing between freshwater (fraction f) and sea water (fraction m) $^{18}\delta$ behaves *conservatively*, that is, $^{18}\delta$ depends only on the mixing ratio of both components, as does the salinity (S) or chlorinity (Cl) (cf. Eq. 8.14):

$$^{18}\delta = \frac{f^{18}\delta_f + m^{18}\delta_m}{f + m} \quad \text{with f} + \text{m} = 1: {}^{18}\delta = f^{18}\delta_f + m^{18}\delta_m \tag{8.18}$$

and

$$S = fS_f + mS_m \quad \text{or} \quad Cl = f\,Cl_f + m\,Cl_m \tag{8.19}$$

$$\text{where} \quad S = 1.80655\,Cl \tag{8.20}$$

A linear relation between $^{18}\delta$ and the salinity can now be obtained by eliminating f and m from these equations (f + m = 1):

$$^{18}\delta = \frac{(S - S_f)^{18}\delta_m - (S - S_m)^{18}\delta_f}{S_m - S_f} \approx 1 - \frac{S}{35}{}^{18}\delta_f = 1 - \frac{Cl}{19.3}{}^{18}\delta_f \tag{8.21}$$

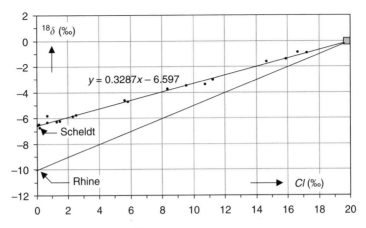

Figure 8.16. Linear relation between $^{18}\delta$ and chlorinity for the estuary of the Western Scheldt, the Netherlands and the estuarine line for the River Rhine (cf. Fig. 3.2).

The salinity of fresh water can be taken as 0‰, while for ocean water $S_m = 35$‰ and $^{18}\delta_m = 0$‰; $^{18}\delta_f$ is more or less specific for each river. Linear mixing is clearly illustrated in Fig. 8.16.

The dependence of $^{18}\delta$ on the salinity is explicitly obtained by differentiating Eq. 8.21:

$$\frac{d^{18}\delta}{dS} = \frac{^{18}\delta_m - ^{18}\delta_f}{S_m - S_f} \approx -^{18}\delta_f/35(\text{‰}/\text{‰}) \tag{8.22}$$

For the North Atlantic a slope of $d\delta/dS = 60$ has been reported, resulting from mixing of the ocean water with fresh polar water with $^{18}\delta \approx -21$‰, presumably representing runoff from the Siberian continent.

In coastal waters receiving multiple river discharge, the origin of the fresh water can sometimes be estimated by using the linearity between Cl and $^{18}\delta$, provided that the rivers have different $^{18}\delta$ values. An example is presented in Fig. 8.16, where $^{18}\delta$ for the River Rhine is about -10‰, and that for the river Scheldt about -6.5‰.

8.2.5 ^{18}O and ^2H in lakes

The oxygen isotopic composition of lake water is primarily determined by that of the river input and, to a lesser degree, by that of direct precipitation and upwelling of groundwater. Changes in $^{18}\delta$ of the water in lakes can occur by evaporation; these aspects have been discussed in detail in Section 8.1.

For proper lake management it is often important to establish a water budget, quantifying all incoming and outgoing water fluxes for the specified time interval. In order to estimate the water balance, stable oxygen and hydrogen isotopes are being applied. The water budget for lakes is based on the mass conservation law and takes the form of a balance equation:

$$\frac{dV}{dt} = I_S + I_G + P - O_S - O_G - E \tag{8.23}$$

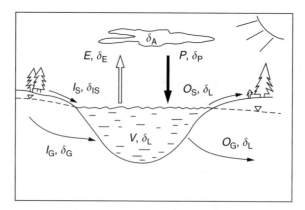

Figure 8.17.   Schematic diagram of the components of hydrological and isotope budgets for a lake. Water fluxes are labelled by capitals, isotopic compositions by corresponding δ values.

See Fig. 8.17: V is the lake volume, I_S, I_G, O_S and O_G represent the volumetric surface and groundwater inflow and outflow fluxes, respectively, P is precipitation over the lake, and E is the evaporation flux from the lake. In principle, all parameters listed in the given equation are functions of time. The water density is assumed constant.

   To derive any given component of the water budget (e.g. the rate of groundwater discharge or recharge and evaporation) from Eq. 8.23, all other parameters either need to be known or can be obtained from independent estimates.

   The tracer approach to establishing a lake water budget is based on the fact that the mass conservation law also applies to any trace constituent built-in within the structure of the water molecule (isotopes of hydrogen or oxygen) or dissolved in the water (e.g. salts). Consequently, the mass balance equation written for the chosen tracer will take the following general form:

$$C_L \frac{dV_L}{dt} + V_L \frac{dC_L}{dt} = C_{IS}I_S + C_{IG}I_G + C_P P - C_{OS}O_S - C_{OG}O_G - C_E E - S$$

$$(8.24)$$

where C with the respective subscripts represents the concentrations of the selected tracer in the lake, as well as in all incoming and outgoing water fluxes entering or leaving the system.

   A condition for the water balance application is that the lake water is well mixed, which may not be the case as is shown by the examples of Fig. 8.18.

   The last term in Eq. 8.23 represents removal of a tracer from the lake by processes other than advection and water fluxes leaving the system. For example, this can be radioactive decay (if a radioactive tracer is applied), a chemical reaction, or adsorption on sedimenting particles. If one has adequate knowledge of tracer concentrations in all functional components of the lake system as functions of time, Eqs 8.22 and 8.23 can be solved numerically for the required pair of variables; a more extensive discussion is presented in chapter 4 of Volume III of the UNESCO/IAEA series.

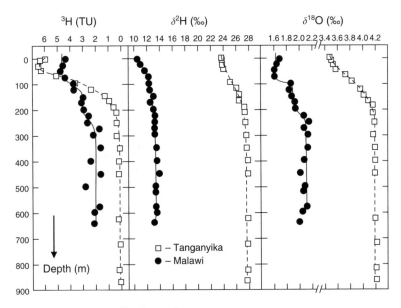

Figure 8.18.   Depth profiles of $^{18}\delta$, $^{2}\delta$ and ^{3}H content in Lake Malawi (1973) and Lake Tanganyika (1975), East Africa. The vertical mixing in Tanganyika lake is extremely slow, as indicated by absence of ^{3}H in the hypolimnion. More positive $^{18}\delta$ and $^{2}\delta$ values at depth, recorded also in Lake Malawi, most probably reflect past climatic changes in the region leading to a reduced ratio of the total inflow to evaporation rate for the studied lakes (see Volume III of the UNESCO/IAEA series).

8.3 ^{3}H IN SURFACE WATER

The majority of applications of the environmental ^{3}H content of surface and groundwaters make use of the mere presence or absence of ^{3}H, or of the ^{3}H concentration peak in precipitation as of 1963 with ^{3}H variations in rain now becoming less distinct relative to the period 1963–70. The possibilities for tracer studies that correlate natural ^{3}H contents in precipitation and surface waters become more and more limited. A more sensitive detection method for ^{3}H is by measuring the presence of ^{3}He, the radioactive daughter of ^{3}H (Section 4.3.4 and Appendix II). This offers further possibilities, for example, of studying the residence age of groundwater.

8.3.1 ^{3}H in rivers and small streams

To a varying extent, the ^{3}H content of river water represents that of precipitation. However, the water may have been subjected to a relatively long residence time in the region, as either groundwater or as surface water (lakes). Especially in larger rivers, the ^{3}H content principally depends on the different ages of the various groundwater components. Moreover, the water may have been transported over a long distance from a region with lower or higher ^{3}H content in the precipitation (Fig. 8.19).

   Figs 8.20 and 8.21 present examples of large and smaller European rivers. It shows that there are two more aspects about the ^{3}H content of rivers, namely the soil often contains,

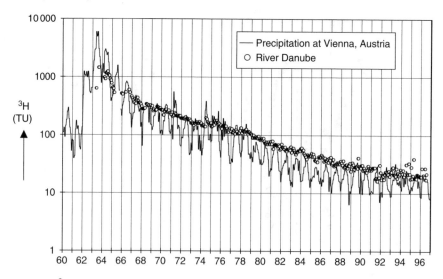

Figure 8.19. ^{3}H in precipitation and in the River Danube at Vienna, Austria (Rank *et al.*, 1998).

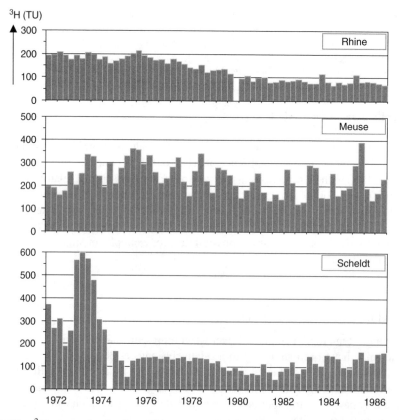

Figure 8.20. ^{3}H content for the rivers Rhine (at Lobith), Meuse (at Eysden) and Scheldt (at Schaar van Ouden Doel), each at the point of entering the national territory of the Netherlands. Each value is the average over a period of 3 months (original data from *Reports on the Quality of Surface Waters in the Netherlands*).

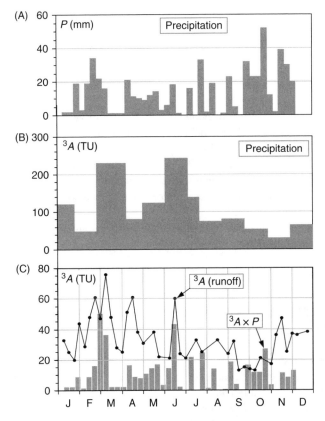

Figure 8.21.   One year of weekly runoff sampling for a small river (Drentse Aa, the Netherlands); (A): weekly precipitation (P), (B): ^{3}H in monthly precipitation, (C): ^{3}H in the runoff; the block diagram in graph (C) indicates the 'amount' of ^{3}H($=^3A \times P$) precipitated.

at least in the Northern hemisphere, ^{3}H produced by or escaped from the testing of nuclear weapons at a level that exceeds the present-day ^{3}H concentration. Furthermore, rivers in certain countries are likely to have, at least periodically, raised ^{3}H levels because of ^{3}H releases by nuclear power stations. The latter possibility is obviously the cause for what happened in the Belgian river Scheldt.

Environmental tritium (^{3}H) and, especially, bomb-^{3}H has shown specific potential to determine the timescale of hydrological processes in the catchment area to be considered.

The ^{3}H content of a small Dutch river over one year varied between about 20 and 60 TU, in reasonable accord with the ^{3}H concentrations in precipitation. ^{3}H input by rain (Fig. 8.21) and that for the stream flow correlate in a similar manner to the ^{18}O response (Fig. 8.14). Especially in summer, the increased rainfall primarily results in an increased discharge of low-^{3}H groundwater.

In principle, the applications of ^{3}H to surface-water studies are similar to those discussed for ^{18}O. The smaller the rivers or streams, the more the runoff represents the ^{3}H content of the regional precipitation, and in the year 2000 in the Northern hemisphere ^{3}H is almost back to the natural levels of around 5 TU.

At present ^{3}H is generally not applicable for studying precipitation-runoff relations, as was shown for ^{18}O. This is due to the fact that the theoretical relation with rain intensity, as shown for ^{18}O, does not exist for ^{3}H (cf. Sections 7.1.3.5 and 8.2.2).

8.3.2 ^{3}H in the sea

The GEOSECS programme has provided a world-wide survey of the ^{3}H content of the Atlantic and Pacific Oceans. The conclusion is that in the open ocean the ^{3}H content is below the detection limit of 0.2 TU at depths exceeding 1000 m, while at the surface ^{3}H concentrations are well below 10 TU. Only in the North Atlantic Ocean does the downward transport of cold water masses cause concentrations of 1–10 TU to be found at a depth of a few kilometres. In these regions the surface water values may extend to 10–20 TU.

In coastal waters, influenced by river discharge, highly irregular ^{3}H patterns are observed with much larger ^{3}H concentrations.

8.3.3 ^{3}H in lakes

The ^{3}H content of lake water is determined by the contributing water components, that is, upwelling groundwater, river water and precipitation. In this context ^{3}H has been applied to studies on: lake-water balance, leakage and vertical mixing.

From a theoretical point of view the applications are simple and straightforward, being based on the linear mixing of ^{3}H as a conservative tracer. The vertical mixing process of a water mass or the downward movement and diffusion, has been extensively studied in the oceans. The oceanic *thermocline*, that is, the transition of mixed layer to stable and low-^{3}H deep water, is likewise observed in lakes as the transition between epilimnion (above thermocline) to hypolimnion (below thermocline). Fig. 8.18 presents clear examples of vertical profiles for ^{3}H as well as ^{18}O and ^{2}H in two African lakes, showing thermocline at about 200 m depth.

8.4 ^{13}C IN SURFACE WATER

The isotopic composition of DIC in surface and also groundwater is closely related to the inorganic carbon chemistry of the water. The solubility of CO_2 in freshwater is in the range of 1.7 to 0.7 L of CO_2/L of water at a pressure of 1 atm and temperatures of 0 and 30°C, respectively. Consequently, the concentration of CO_2 at the pre-industrial atmospheric CO_2 pressure of 280×10^{-6} atm (presently 380 ppm) is in the range of 0.5–0.2 mL/L of water, equivalent to about 0.02–0.01 mM of carbon/L of water (mM/L). The natural concentration in fresh water is often 2–5 mM of carbon/L. The origin of carbon dissolved in natural waters is thus not simple dissolution of atmospheric CO_2. This fact is confirmed by the isotopic composition, as will be shown. Because the inorganic carbon chemistry is essential for understanding as well as applying the carbon isotopic composition of dissolved carbon, we include a brief presentation of the main chemical facts and equations for the carbonic acid equilibria in Appendix III, illustrated by a few examples.

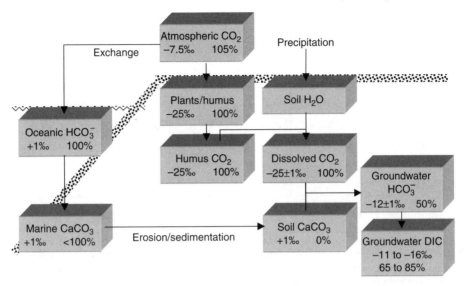

Figure 8.22. Combination of Figs 5.6 and 6.4, describing the ^{13}C isotope ratio as $^{13}\delta_{PDB}$ and the ^{14}C content as ^{14}a of groundwater.

8.4.1 The origin of inorganic carbon in natural waters

Since most natural waters, apart from meteoric water, are or have once been groundwater, most DIC has been developed according to the schemes presented in Sections 5.1.4.6 and 6.1.2.4.

In general, the isotopic composition of dissolved HCO_3^-, by far the largest carbon fraction in groundwater and derived surface waters, is the most meaningful as far as the origin is concerned (Fig. 8.22). Therefore, we will generally discuss $^{13}\delta(HCO_3^-)$.

8.4.2 ^{13}C in rivers and small streams

As stated earlier, the observation that large rivers in general contain groundwater is confirmed by the isotopic composition of the DIC. In winter, rivers have $^{13}\delta$ values of dissolved bicarbonate (the largest fraction) which are typical for groundwaters (Fig. 8.23). The marked seasonal variation results from $^{13}\delta$ changes during summer. Isotopic exchange between dissolved carbon and atmospheric CO_2 is responsible for this change. Exchange sufficient to bring about the observed $^{13}\delta$ changes should also cause chemical changes such as increasing pH by the loss of CO_2. Although such changes have been observed in lakes (Section 8.4.3), they are not found in the large rivers to the extent expected. This can be explained by assuming that the isotopic exchange does not take place in the river, but rather in the smaller, shallow streams feeding the river. Because of the relatively large surface area, the $^{13}\delta$ values and pH can increase significantly. When this (partly) exchanged water mixes with normal groundwater in the river, $^{13}\delta$ simply obeys linear mixing. The relatively low pH, however, is only slightly influenced by the high pH of the exchanged water. A calculation shows that if two equal quantities of water with the same alkalinity (3 mM/L) and pH values of 7 and 9, respectively, are mixed, the resulting pH of the mixture is 7.35. The reasoning is

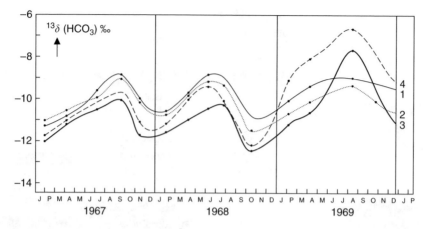

Figure 8.23.   Seasonal variations of $^{13}\delta$ for dissolved bicarbonate in the main rivers in the Netherlands. The values in winter illustrate the groundwater origin of the river water. The fall of 1968 was extremely wet, which caused the $^{13}\delta$ decline to be relatively early in the year. The summer of 1969 was extremely dry and hot, affecting the smaller rivers in particular.

supported by the high $^{13}\delta$ during the drought in North Western Europe in 1969, as opposed to the extremely wet fall of 1968.

Small rivers and streams show much more pronounced and irregular variations in $^{13}\delta$, partly correlated with the precipitation intensity. During heavy rainfall, groundwater with $^{13}\delta \approx -12‰$ is forced into the river channel, while the rapid discharge of the surface runoff barely allows any isotopic exchange with the atmosphere and in any case, contains little carbon.

8.4.3 ^{13}C in lakes

This section on lakes discusses what happens to the isotopic composition of dissolved carbon when the water is exposed to the air for a prolonged time. These processes are observed in any type of water mass, small or large, in rivers and estuaries, and canals, pools and in irrigation water before infiltration.

In the first example the carbon isotopic composition of lake water is determined by that of the principal feeding components, that is, river water and groundwater. Precipitation hardly plays any role, because the carbon content of rain is very low. Considering the origin of the water, $^{13}\delta$ of the dissolved bicarbonate is generally in the range of -11 to $-12‰$, depending slightly on the season.

Carbon isotopic changes in a lake are brought about by the fact that during the residence time of the water in the basin it is exposed to the air and to certain chemical and biological processes in the lake water. We can distinguish four different processes which are chemically linked to form an often complicated picture:

1.   Chemical equilibration with atmospheric P_{CO_2}.
2.   Precipitation of $CaCO_3$.

3. CO_2 consumption by biological activity.
4. Isotopic exchange with atmospheric CO_2.

8.4.3.1 *Chemical equilibration with atmospheric P_{CO_2}*

Due to large partial-CO_2 pressure in the unsaturated soil zone, ground- and river water correspond to P_{CO_2} values of $10^{-3}-10^{-1}$ atm. If this water is exposed to the air, the dissolved CO_2 concentration decreases to reach equilibrium with the atmospheric CO_2 concentration at $P_{CO_2} \approx 3 \times 10^{-4}$ atm. This process causes the isotopic composition of the dissolved carbon to change correspondingly. The isotopic change can be calculated by applying results from the model presented in Section III.5.2.1 in Appendix III. In most groundwater and derived waters the fraction of CO_2 is in the order of 10% (pH ≈ 7.0–7.5); exposed to the atmosphere, the water essentially loses the dissolved CO_2 fraction. The resulting change in $^{13}\delta$ is then (cf. Eq. 3.8):

$$\Delta^{13}\delta = \left(\frac{C}{C_0}\right)^{\varepsilon} - 1 \qquad (8.25)$$

With a loss in dissolved CO_2 of 10% of the DIC (primarily HCO_3^-) and $\varepsilon_{g/b} = -9.54\%o$, the change is $(0.9)^{-0.00954} = +1.0\%o$.

8.4.3.2 *Precipitation of calcium carbonate*

The escape of CO_2 from the water gives rise to a metastable state in which the solution is oversaturated with respect to $CaCO_3$. In due course calcium carbonate is likely to precipitate according to:

$$2HCO_3^- \leftrightarrow CO_3^{2-} + CO_2 + H_2O$$
$$Ca^{2+} + CO_3^{2-} \leftrightarrow CaCO_3 \qquad (8.26)$$

The formed CO_2 will escape from the water and be at a sustained equilibrium with the atmospheric P_{CO_2}. The precipitation process does not affect the pH of the solution. Both the formation of carbonate and the additional loss of CO_2 cause changes in $^{13}\delta(HCO_3^-)$. This process is described in Section 3.2.2 (Fig. 3.5, Eq. 3.12) by a change in $^{13}\delta$ of:

$$\Delta^{13}\delta = \left(\frac{C}{C_0}\right)^{f_1\varepsilon_1 + f_2\varepsilon_2} - 1 \qquad (8.27)$$

where f_1 and f_2 are the fractions of $CaCO_3$ and CO_2 removed from the water; $f_1 = f_2$ since the precipitation of $CaCO_3$ and escape of CO_2 are coupled processes. C and C_0 refer to the actual and the original bicarbonate concentrations, respectively; $\varepsilon_1 = \varepsilon_{s/b}$ and $\varepsilon_2 = \varepsilon_{g/b}$. The effective fractionation (exponent in Eq. 8.25) at 20 °C is: $\varepsilon_{eff} = 0.5[-0.00846 + 0.00066] = -3.9\%o$ (see Table 5.2). For example, a realistic alkalinity drop from about 2.8 to 1.8 mM/L at 20°C results in an increase in $^{13}\delta$ of $+1.7\%o$.

8.4.3.3 *CO_2 consumption by biological activity*

During growth of algae CO_2 is consumed which is much depleted in ^{13}C with respect to the bicarbonate. This fractionation is deduced from a series of observations: $^{13}\varepsilon_b$ (algae) $\approx -23\%o$. The uptake of CO_2 by the algae results in a pH change and, subsequently, precipitation of $CaCO_3$. Again we can apply Eq. 8.27 with $f_1 = f_2$ and ε_1 and ε_2. In this case the net fractionation effect is $\varepsilon_{\text{eff}} = 0.5[-23 + 0.00066]$. If the observed decrease in the total carbon content in the lake from 3.0 to 2.0 mM/L would originate entirely from the consumption of 0.5 mM/L of CO_2 by algae and a simultaneous precipitation of 0.5 mM/L of carbonate, the resulting isotope effect at 20 °C would be about +4.5‰.

8.4.3.4   *Isotopic exchange with atmospheric CO_2*

Finally, $^{13}\delta$ of the dissolved carbon changes by isotopic exchange with atmospheric carbon dioxide so that $^{13}\delta$ 'moves' in the direction of an isotopic equilibrium with the atmospheric CO_2, but not necessarily reaching isotopic equilibrium. This depends on the time available, because it is a relatively slow process, the isotopic change having to be distributed over the entire water column. Here the degree of turbulence of the water is the most important factor determining the exchange rate. This varies between 10 and 100 moles of carbon/m^2 per year (Fig. 8.24). In principle, the degree of ^{13}C exchange may offer the possibility for estimating the average residence time of water in a lake or estuary.

8.4.4 ^{13}C in the sea

The $^{13}\delta$ value of ocean water bicarbonate varies between +1 and +2.5‰, with the majority of data between +1.5 and +2‰. The corresponding range of $^{13}\delta$ in total DIC is about +1.0 to +2.0‰.

A condition for isotopic equilibrium with atmospheric CO_2 with $^{13}\delta \approx -7.5$‰, requires sea water HCO_3^- to be about +2‰ at 10 °C and +1‰ at 20 °C, using the fractionation factors in Table 5.2.

The isotopic composition is barely affected by seasonal changes in temperature or in $^{13}\delta$ of atmospheric CO_2, because of the relatively low exchange rate. However, $^{13}\delta$ variations are caused by biological activity. Growth of algae involves a fractionation of about -20 to -23‰, so that a consumption of 5% of C_T already results in a $^{13}\delta$ change of $-0.05 \times (-20$‰$) = +1$‰.

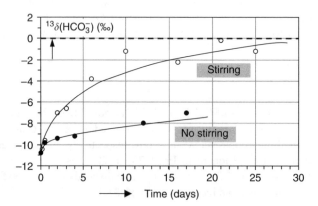

Figure 8.24.   Result of a simple experiment which exchanges regular tap water with the air-CO_2 with $^{13}\delta$ in the range of -8 to -8.5‰ at 20 °C; $^{13}\delta(HCO_3^-)$ changes towards isotopic equilibrium with air-CO_2: -8.5‰ $+^{13} \varepsilon_{b/g} \approx 0$‰. The influence of stirring on the exchange rate is obvious.

8.4.5 ^{13}C in estuaries

The $^{13}\delta$ value for the total DIC in an estuary is determined by the mixing ratio of river and sea water, as are $^{18}\delta$ and the salinity or chlorinity (Fig. 8.16). This is only true if the total dissolved carbon as well as ^{13}C are conservative; that is, no production or consumption of carbon, and no isotopic exchange occurs. This does not result in a linear relation between $^{13}\delta_T$ and S (or Cl) as is observed between $^{18}\delta$ and S (Eq. 8.16). The reason is that, in general, the C_T values for the fresh and sea water are not equal. With f and m referring to the fractions of fresh and sea water and $f + m = 1$, the ^{13}C mass balance can be written in terms of $^{18}\delta$ of the water. From:

$$C_T{}^{13}\delta_C = fC_{Tf}{}^{13}\delta_{Cf} + mC_{Tm}{}^{13}\delta_{Cm}$$
$$\text{and} \quad {}^{18}\delta = f{}^{18}\delta_f + m{}^{18}\delta_m \qquad \text{(cf. Eq. 8.18)} \tag{8.28}$$

the ^{13}C mass balance is:

$$^{13}\delta_T = \frac{(C_{Tf}{}^{13}\delta_f - C_{Tm}{}^{13}\delta_m)^{18}\delta + C_{Tm}{}^{13}\delta_m{}^{18}\delta_f - C_{Tf}{}^{13}\delta_f{}^{18}\delta_m}{(C_{Tf} - C_{Tm})^{18}\delta + C_{Tm}{}^{18}\delta_f - C_{Tf}{}^{18}\delta_m} \tag{8.29}$$

The good agreement between the measured and calculated curve for the Western Scheldt estuary, the Netherlands, is shown in Fig. 8.25.

The conservative character of $^{13}\delta_T$ only applies to those cases, where no isotopic exchange or other processes affecting the ^{13}C content occur; that is, if the residence time of the water is relatively short and the water is not shallow.

Fig. 8.26 presents an overview of the $(^{18}\delta, {}^{13}\delta)$ relations for the various categories of surface waters discussed in this chapter. The following is a summary of the

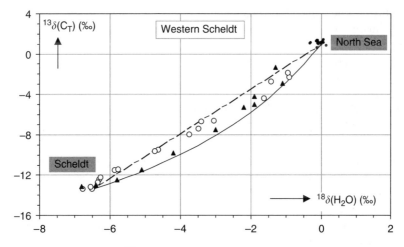

Figure 8.25. Relation between $^{13}\delta$ of DIC in the estuary of the Western Scheldt (Belgium and the Netherlands) and $^{18}\delta$ of the water, representing the mixing ratio of the fresh- and the sea water. The full line is calculated according to Eq. 8.29.

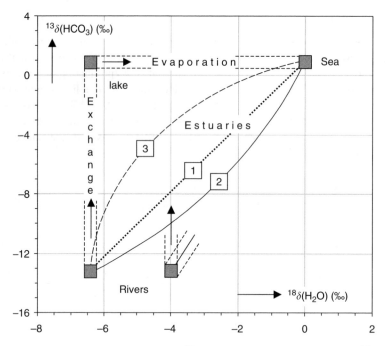

Figure 8.26.   Diagram of the relation between $^{13}\delta$ for dissolved (bi)carbon(ate) and $^{18}\delta$ of natural waters.

different processes:

- $^{13}\delta$ of river DIC is based on groundwater values.
- $^{18}\delta$ of river water depends on precipitation in the catchment.
- The estuarine ($^{13}\delta$, $^{18}\delta$) relation depends on the total carbon content (C_T) of the components.
- If $C_{Tf} = C_{Tm}$ (= 2 mM/L), the estuarine line is straight (line 1).
- Line 2 represents higher values of C_{Tf}, line 3 lower values.
- Evaporation causes an increase of $^{18}\delta(H_2O)$.
- Isotopic exchange with air-CO_2 causes an increase of $^{13}\delta(HCO_3)$.
- Line 3 is likely to represent exchange with air-CO_2 in estuaries.

A second overview (Fig. 8.27) deals not only with *Dissolved Inorganic Carbon* (DIC) in the water, but also with the *Particulate Inorganic Carbon* (PIC $\equiv CaCO_3$ essentially) and the *Particulate Organic Carbon* (POC).

The organic matter may be of continental, riverine/estuarine or marine origin. The continental origin is revealed, in particular, by a relatively high age and thus low ^{14}a. The $CaCO_3$ fraction is generally of continental origin, again manifest by low ^{14}C. The $^{13}\delta$ value for organic matter is also determined by the origin, taking into account that the fractionation of plankton relative to dissolved HCO_3^- is about $-23‰$. Consequently, $^{13}\delta$ for marine POC is $+1 - 23 = -22‰$, while riverine POC has a $^{13}\delta$ of $-12 - 23 = -35‰$, contrary to continental POC – primarily eroded peat – with $^{13}\delta \approx -26 \pm 1‰$.

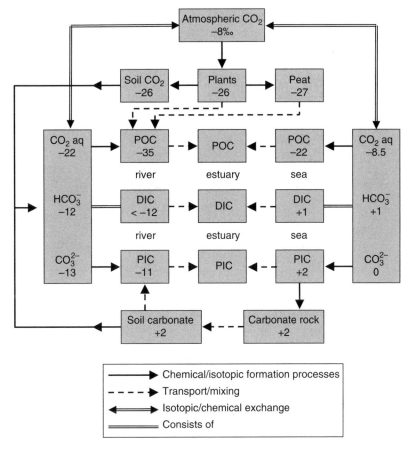

Figure 8.27.   Schematic diagram of the sources of DIC, suspended POC and PIC (primarily CaCO$_3$) in rivers, estuaries and the sea. The $^{13}\delta$ numbers (in ‰) are representative averages. The carbonic acid and carbonate isotope fractionation values are given in Chapter 5.

8.5 ^{14}C IN SURFACE WATER

When studying the ^{14}C abundance in surface waters, isotopic effects play only a minor role. The ^{14}C content of the inorganic carbon is primarily determined by the ages of the contributing water components, primarily groundwater. Isotopic exchange, however, can significantly alter the ^{14}C content of surface water, proportional to the ^{13}C change. This effect has to be taken into account, for example, when dating lake sediments.

CHAPTER 9

Isotopes in groundwater

Isotopic methods applied to groundwater studies are primarily aimed at the following questions:

- *What is the origin of the groundwater?*
  Here we consider problems such as: where is the recharge area of the aquifer to be found; does the water originate from direct infiltration of precipitation or (partly) from a surface-water body (river, lake or sea).
- *How does the groundwater move?*
  This question concerns the flow direction and flow velocity of a groundwater body, as well as the age of a stagnant water mass. Moreover, isotope studies often reveal local infiltration rates.
- *Which geochemical processes operate underground?*
  Processes such as the anaerobic decomposition of organic matter (peat) and exchange with clay minerals or carbonate rock affect the chemical as well as the isotopic composition of the groundwater. Isotopic evidence might allow conclusions about chemical changes, and *vice versa*. The latter is especially relevant, because in those cases the isotopic composition of the groundwater no longer reflects the infiltrated water or the age.

As we have seen in the preceding chapters, each isotope presents a specific part of the overall picture and will therefore be treated separately. Carbon isotopes are not constituents of the water molecule and, consequently, pose many complications and deserve the most attention. This is the more justified, since groundwater dating is important in relation to the exploration of drinking water supplies, as well as problems of underground disposal of nuclear waste.

9.1 ^{18}O IN GROUNDWATER

9.1.1 Correlation between ^{18}O of groundwater and precipitation

The possibility of determining the origin of groundwater from the $^{18}\delta$ value implies knowledge about the $^{18}\delta$ of precipitation. Without complications, the $^{18}\delta$ of groundwater represents that of the precipitation during periods of the largest infiltration. We will return to this problem in more detail in our discussion of the ^{3}H data (Section 9.3). In regions with temperate climatic conditions the observed $^{18}\delta$ values are in reasonable agreement with those for the average yearly rainfall. Fig. 9.1 shows a histogram of $^{18}\delta$ values observed in Dutch groundwater samples, related to $^{18}\delta$ of precipitation; most samples reflect the average

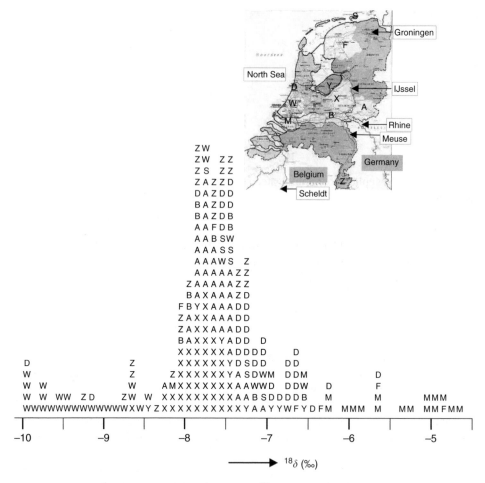

Figure 9.1.   Histogram showing the occurrence of $^{18}\delta$ values in groundwater samples from the Netherlands (see map for the groundwater sites and the rivers). $^{18}\delta$ of annual precipitation (represented by Groningen) is in the range of $-8\pm0.5\permil$; $^{18}\delta$ of Rhine water is $-10\permil$. A: Achterhoek; B: Betuwe; D: Dunes, Western Netherlands; F: Friesland; M: Saline water; S: Schiermonnikoog; W: Polders, Western Netherlands; X: Veluwe; Y: IJsselmeerpolders; Z: Southern Netherlands.

rain over many years. Due to a small *continent effect*, the $^{18}\delta$ values along the North Sea coast are slightly higher (D in Fig. 9.1). Several samples from the region of the river delta (W in Fig. 9.1) obviously originated from infiltrated river water (cf. Rhine in Fig. 8.13). The same is true for some samples collected around the artificial infiltration area for drinking water supply in the dune area (D).

   In (semi-)arid regions the correlation between $^{18}\delta$ in groundwater and precipitation is less simple, since groundwater recharge only occurs during the few periods of high precipitation intensity. Light rainfall will completely evaporate before infiltration can take place. Although in these regions the $^{18}\delta$ variations in rain may be very large, the $^{18}\delta$ values observed in groundwater are surprisingly small. In some cases $^{18}\delta$ has increased by evaporation before infiltration; this clearly follows from the ($^{18}\delta$, $^2\delta$) relation (Fig. 8.10).

9.1.2 ^{18}O in saline water

Another group of groundwater samples, marked M in Fig. 9.1 consists of waters which partly contain fossil or non-fossil seawater, explaining the relatively high $^{18}\delta$ values. Here the chlorinity is correlated with $^{18}\delta$ according to (cf. Section 8.2.4):

$$^{18}\delta = \frac{S_m - S}{S_m}{}^{18}\delta_f \tag{9.1}$$

(where $S_m = 35\%o$, if desired to be replaced by $Cl_m = 19.4\%o$ and S_m is taken $= 0\%o$), pointing to mixtures of freshwater ($^{18}\delta_f$) and seawater ($^{18}\delta_m = 0\%o$).

A few samples with high S values (Fig. 9.1) have relatively low $^{18}\delta$ values. These concern originally fresh groundwaters which contain a high chloride content because of water–rock interaction (salt dissolution), not affecting the $^{18}\delta$ values. In semi-arid regions a high salt content in groundwater samples is not necessarily related to a marine origin. The water might also have been subjected to evaporation, for example, in open wells. A conclusion about the one or the other cause should be drawn from the ($^{18}\delta$, $^2\delta$) relation (Section 9.2).

In principle, infiltration rates can be determined by correlating $^{18}\delta$ of soil water, collected by lysimeters or mini-filters, with $^{18}\delta$ in precipitation. This application of ^{18}O variations is in reality hard to perform, since the precipitation has to be sampled very frequently and regularly, while suitable ^{18}O tracer peaks are rare (cf. Fig. 8.15).

In groundwater studies, ^{18}O analyses are generally performed in relatively large quantities in order to establish the homogeneity of a groundwater mass. If $^{18}\delta$ turns out to be inhomogeneous, it is less useful to compare ^3H or ^{14}C ages of samples from one aquifer.

9.2 ^{2}H IN GROUNDWATER

It is a common practice to analyse both ^{18}O and ^2H in groundwater samples in those cases where $^{18}\delta$ shows a marked deviation from a uniform value. The ($^{18}\delta$, $^2\delta$) relation will demonstrate whether the sample originates from direct infiltration of meteoric water; in temperate climates this is nearly always true.

Groundwater can have a high salt content by underground mixing with seawater, by evaporation before infiltration or by underground dissolution of salt. Apart from possible chemical evidence in this respect, the ($^{18}\delta$, $^2\delta$) relation might also reveal the history of the water. This is shown in Fig. 8.10. In the third case the ($^{18}\delta$, $^2\delta$) values will coincide with the meteoric water line.

Transpiration – the most common type of evaporation in regions of moderate climates, that is, through the plant leaves – does not cause significant isotopic enrichment in the soil water. This phenomenon may be compared with the case shown in Fig. 3.7. The constant supply of soil water through the capillaries to the leaves causes isotopic enrichment in the leaves, as the water vapour released by the leaves will have the same isotopic composition as the soil water.

In a few cases, deviating (high) $^{18}\delta$ values in thermal groundwater have been explained by isotopic exchange with carbonate rock. This exchange process is extremely slow at normal temperatures, but is strongly increased at high temperatures. The ^2H content of the water is not affected by this water–rock interaction (Fig. 8.10).

9.3 ^{3}H IN GROUNDWATER

9.3.1   Applications

^{3}H analyses are commonly applied in groundwater research because:

- For 'young' groundwater (<50 years), ^3A can within limits be used for age determination.
- For 'old' groundwater (>100 years), the ^3A value shows that the water sample, intended for ^{14}C age determination, does not contain a recent component.
- ^3A variations in precipitation can be used as a tracer for measuring a local infiltration rate.
- ^{3}H analysis can be applied to differentiate between water infiltration or upwelling.

9.3.2 ^{3}H in infiltration studies

Here we will distinguish between the *infiltration velocity* and the *infiltration rate*. In isotope hydrology one can make this distinction, because water is characterised by the isotopic composition: 'pure water can be distinguished from pure water'. With isotopes as ideal tracers, the water movement can be so determined. In classical hydrology natural or artificial chemical tracers are sometimes used. The disadvantage is that the chemical compounds are often adsorbed to the soil constituents and, consequently, they tend to show apparent water velocities which are smaller than the true velocities.

9.3.2.1   *Determination of the infiltration velocity with ^{3}H*
In Section 7.2.2.2 we discussed the phenomenon that the injection of ^{3}H into the atmosphere during the early 1960s was not evenly distributed over the seasons. Figs 7.17 and 9.2 show this seasonal effect.

The consequence is that with the downward displacement of rain in the soil the ^{3}H distribution can be recognised as a peak. Comparison of the ^{3}H distribution in the soil with that in the precipitation (October–March) can, therefore, result in an estimate of the infiltration velocity of the water. With an interval of one year, samples were collected from a boring containing 14 filters spaced at 1 metre. From a hydrological point of view, the water in this dune area is expected to infiltrate in a purely vertical direction. The phenomenon of dispersion – that is, broadening of the 1963 ^3A peak – restricts the precision of the determination. Nevertheless, a velocity of 1 m/year following from the peak depth in 1977 is in reasonable agreement with the vertical shift over the period 1977–78. If we assume the absence of a horizontal component and a porosity of 0.3, it appears that about 300 mm of rain infiltrates per year, about 40% of the total rainfall. The remaining 60% is withdrawn from the soil by evaporation or evapotranspiration.

9.3.2.2   *Determination of the infiltration rate with ^{3}H*
The infiltration or recharge rate, $R(t)$, can be determined by comparing the known amount of infiltrated ^{3}H, ^3A_i, with the time-dependent ^{3}H content of the precipitation, $^3A(t)$:

$$^3A_i = \int_{t_1}^{t_2} R(t)^3A(t)dt \quad \text{(TU metres)} \tag{9.2}$$

$^3A(t)$ is given by Figs 8.16 and 8.17 or Fig. 9.3.

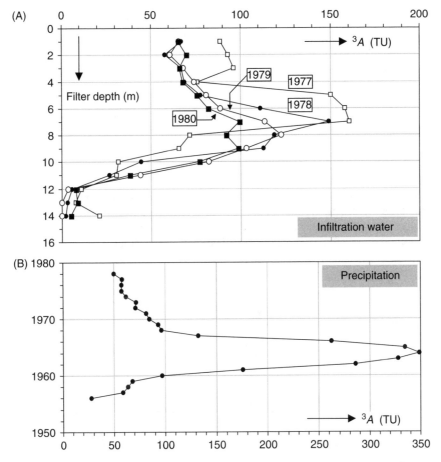

Figure 9.2. (A) ^{3}H profile of infiltrated water in a dune area in North Western Europe. (B) ^3A in winter precipitation (October–March) average for Vienna and Ottawa; a dispersion has been introduced equivalent to a 5-year moving average (all data corrected for radioactive decay to 1980) (cf. Fig. 9.10).

^3A_i has to be determined by integrating the product of the ^{3}H content as a function of depth, $^3A(x)$, and the amount of water $W(x)$ that is contained by the soil zone at issue to a depth d:

$$^3A_i = \int_0^d W(x)\,^3A(x)\,\mathrm{d}x \tag{9.3}$$

^3A_i can also be deduced from the water collected by a lysimeter. Generally, Eq. 9.2 will be more practical in discrete form, because the $^3A(t)$ values are known per month or per year:

$$^3A_i = \sum_i R(t)^3A(t) = <R> \sum_i {}^3A(t) \tag{9.4}$$

where $R(t)$ is the infiltration per month or year during the period of the experiment. In this way the average infiltration rate, $< R >$, can be found:

$$< R > = \frac{^3A_i}{\sum_i {}^3A(t)} \tag{9.5}$$

where $\Sigma^3A(t)$ is the integral or surface area of a specific section of the ^{3}H distributions mentioned earlier.

9.3.3   Dating groundwater with ^{3}H

As mentioned in the introduction, the natural ^{3}H content of precipitation was completely overshadowed by thermonuclear ^{3}H brought into the atmosphere by the test explosions of especially hydrogen bombs in the early 1960s. Even an admixture of 1% of rain water from this period (Section 7.2.2) completely disturbed the natural ^{3}H level.

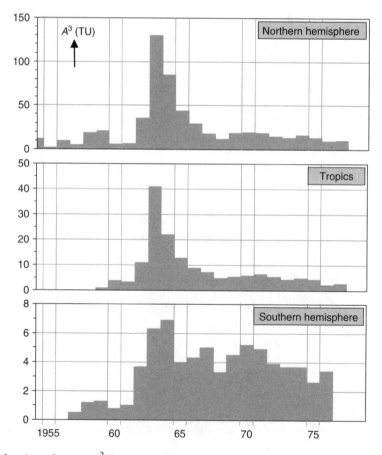

Figure 9.3.   Annual average ^{3}H content of precipitation representative for the Northern and the Southern hemispheres, and the tropics. The data are taken from the IAEA/WMO network and are corrected for radioactive decay till the year 2000.

Because of the uncertainty of the input ^{3}H concentration, we can not apply ^3A for absolute dating water *based on the radioactive decay process*, as with ^{14}C.

On the basis of Fig. 9.3, dating of precipitation from one year would be possible, though this is obviously unrealistic. We have to compare the ^3A of the groundwater sample with the infiltrated ^{3}H over the entire infiltration period. However, this is the unknown in our problem and leaves us with two possibilities:

1. *Recovering the ^3A 'spectrum' of precipitation in the water body (in some cases).* In general, the ^{3}H 'spectrum' is heavily disturbed because of a combination of vertical and horizontal flow patterns. Only in a very local search for infiltration characteristics could we have a chance for success; an example was given in Section 9.3.2.1.
2. *Putting limits to the water age.* Here we can make an approach (literally) from two opposite directions:

    (a) Sub-recent water dating prior to 1950 has a ^{3}H content below 1 TU (more than two half-lives old). This can have become mixed in the aquifer with more recent water with ^3A up to 120 TU (Fig. 9.3). The observed ^3A value therefore sets a limit to the percentage of recent water. Additional hydrological evidence is thus needed to decide what is the most likely. For each hemisphere it is possible to approximately indicate the average ^3A value of a water mass including a certain number of years from the 'thermonuclear period'. We restrict ourselves here to illustrate this point: the actual data have to be taken from a large number of weather stations and do not take into account the geographically varying precipitation intensities. It is obvious, however, that ^3A values exceeding, for example, 200 TU point to water dating from a short period around 1963. From Fig. 9.10 we read two possibilities for explaining ^3A of 45 TU: water from 1950–62 or from 1970–80. It is evident that in reality there are more possibilities, which need to be related to other hydrological information.
    (b) In Fig. 9.10 we also indicated the ^{3}H content for water originating from a period prior to the present. From the curve it appears that an ^3A value above 200 TU is not to be expected. Here we again assumed that the infiltration is not evenly spread over the year; we will return to this in Section 9.7 (Fig. 9.10).

More appropriate curves have to be calculated from weighted average ^3A values over the most likely periods of infiltration. These are obtained by multiplying the respective input $^3A(t)$ (Chapter 7; WMO/IAEA data) with the estimated amount of infiltration, or the precipitation in the project region.

The purpose of the foregoing discussion is to present a way of thinking about ^{3}H values, rather than deriving a generally valid manual for age determinations.

9.3.4   Absolute dating of groundwater with ^{3}H/^{3}He

The application of ^{3}H to obtain underground residence times ('ages') of groundwater is stated in Section 9.3.3 to depend on our knowledge of the ^{3}H input from atmospheric precipitation. Studying the relation between ^{3}H and ^{3}He offers the possibility to determine ages without knowing the tritium input function. If the original ^{3}H content of the infiltrating precipitation is $^3A_0$, ^3A after a residence time t is:

$$^3A_t = {}^3A_0 \, e^{-\lambda t} \qquad\qquad (9.6)$$

The amount of ^{3}He formed during this period of time is equal to the amount of ^{3}H lost by decay (Section 4.3.4):

$$^3He_t = {}^3A_0 - {}^3A_t = {}^3A_t(e^{\lambda t} - 1) \qquad\qquad (9.7)$$

From Eqs 9.6 and 9.7 the unknown $^3A_0$ can be eliminated, resulting in the ^{3}H/^{3}He age:

$$t = (1/\lambda) \times \ln[{}^3He_t/{}^3A_t + 1] \qquad\qquad (9.8)$$

(see actual data in Section 4.3.4). Most applications are restricted to shallow, young groundwater, using the bomb peak of ^{3}H in precipitation during the early 1960s.

A drawback of the method is that the occurrence of ^{3}H and ^{3}He at great depth may be due to fractures, cracks or leaking boreholes, or to *in situ* production of the gases.

^{3}He may be of non-tritiogenic or terrigenic origin, that is, from *in situ* production, according to

$$^6Li + n \rightarrow {}^3He + {}^4He \quad \text{or} \quad {}^6Li(n, \alpha){}^3He \qquad\qquad (9.9)$$

where the underground neutrons originate from spontaneous fission of Uranium and Thorium isotopes. Part of the ^{3}He may come from the original equilibration of the water with atmospheric gases. The latter contribution is estimated from the Ne concentration, which cannot be of terrigenic origin. A possible terrigenic contribution is estimated from the concentration of ^{4}He (also corrected for the atmospheric component) details are given in Volume I of the UNESCO/IAEA series.

> The main consequence of a possible terrigenic production of ^{3}He is that the method is primarily applicable in shallow groundwater.

In these applications the ^{3}H activity is very low. Additionally, the required precision in ^3A analysis is high. The solution to these contradictory conditions is the so-called *^{3}He-ingrowth technique* (Appendix II).

9.3.5   Infiltration versus upwelling

From ^{3}H measurements of water extracted from strings of mini-filters, one can easily distinguish between the occurrence of infiltration or upwelling water. Results are shown in Fig. 9.4 for an area where precipitation is locally infiltrating or groundwater is upwelling and drained off by a small stream.

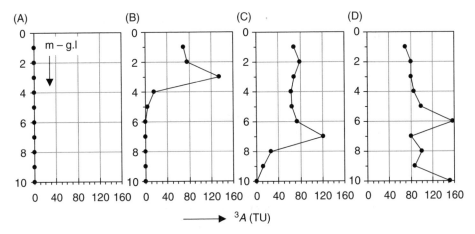

Figure 9.4. ^{3}H profiles of shallow groundwater in an area where water infiltration as well as upwelling may occur, as shown by the presence, respectively, absence of ^{3}H in the water samples. The samples were taken from mini-filters at vertical distances of 1 m (metres minus ground level) (data valid for the year of sampling = 1980). The samples were collected from the Barneveldse Beek drainage basin (central part of the Netherlands); sample A: obviously no infiltration, but rather upwelling; B–C: positive indication of infiltration, down to increasing but limited depths.

9.4 ^{13}C AND ^{14}C IN GROUNDWATER

Dating groundwater with ^{14}C is the important issue in isotope hydrology, because there is much need for a dating technique in this particular age range. On the other hand, several aspects of the principles remain in dispute. Causes for this are:

- The dissolved inorganic carbon, by which compound the ^{14}C is contained, can be of different origin.
- Because the carbon content in groundwater is small, the ^{13}C and ^{14}C content can more easily be disturbed by soil processes than the other isotopes.
- Several models are circulating which are based on different and partly contradictory starting points, and (accidentally) which seem to specifically apply to only the hydrogeological systems studied by the presenting authors.

The confusion can be reduced to two questions:

1. What is the original ^{14}C content of the recently infiltrated water?
2. Has the ^{14}C content been subsequently altered in the aquifer by chemical or biological processes?

The answer to the first question depends primarily on the origin of the dissolved inorganic carbon. In the next section we discuss possible carbon sources and their resulting isotopic contents. The symbols used are compiled in Table 9.1 and Fig. 9.5.

Table 9.1.   Notation used in this chapter for the concentration of the carbonic acid fractions and the ^{13}C and ^{14}C content; some of the symbols are referred in Fig. 9.5.

a, b, c	(Molar concentrations of) dissolved CO_2, HCO_3^- and CO_3^{2-}, respectively, generally in mM/L (millimoles/litre)
Σ	Molar concentration of total dissolved inorganic carbon (DIC) $= a + b + c$
g	Gaseous CO_2
$^{14}a_{go}$, $^{13}\delta_{go}$	^{14}C (relative) activity (in %) and ^{13}C content of gaseous soil CO_2
$^{14}a_{ao}$, $^{13}\delta_{ao}$	idem of dissolved CO_2
$^{14}a_{lo}$, $^{13}\delta_{lo}$	idem of marine $CaCO_3$
$^{14}a_l$, $^{13}\delta_l$	idem of actual soil $CaCO_3$
$^{14}a_{bo}$, $^{13}\delta_{bo}$	idem of dissolved HCO_3^- prior to exchange
$^{14}a_{bu}$, $^{13}\delta_{bu}$	idem of dissolved HCO_3^- after partial exchange with gas 'go'
$^{14}a_{be}$, $^{13}\delta_e$	idem of dissolved HCO_3^- after complete exchange
$^{14}a_{bs}$, $^{13}\delta_{bs}$	idem of actual (i.e. measured) dissolved HCO_3^-
$^{14}a_{\Sigma}$, $^{13}\delta_{\Sigma}$	idem of DIC in recent recharge in the area
$^{14}a_m$, $^{13}\delta_m$	idem of DIC in the actual water sample
$^{13}\varepsilon_g$, $^{13}\varepsilon_a$, $^{13}\varepsilon_c$	^{13}C fractionation (in ‰) of gaseous CO_2, dissolved CO_2 and CO_3^{2-} relative to dissolved HCO_3^- (resp. $^{13}\varepsilon_{g/b}$, $^{13}\varepsilon_{a/b}$, $^{13}\varepsilon_{c/b}$ in Table 5.2).

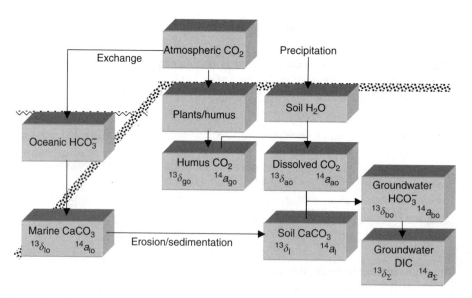

Figure 9.5.   Part of the carbon cycle explaining the isotopic composition of DIC in groundwater. The symbols refer to the definitions in Table 8.1 (figure also shown in Chapters 5, 6 and 8).

9.4.1   The origin of ^{13}C and ^{14}C in groundwater

Several sources are presented in literature for the dissolved inorganic carbon content. To each carbon source we devote a separate section. The first group of four processes occur in the upper soil layers, mostly in the unsaturated zone.

9.4.1.1   *Oxidation of organic matter in the upper soil*

CO_2 produced by the decomposition of organic material in the soil (humus), or by root respiration in the unsaturated zone can dissolve soil carbonates:

$$CaCO_3 + CO_2 + H_2O \rightarrow Ca^{2+} + 2HCO_3^- \qquad (9.10)$$

Starting from a ^{14}C content in the organics of 100% (natural recent ^{14}C level) and in the limestone of 0% (old marine carbonates), the dissolved bicarbonate is expected to have an original ^{14}C content $^{14}a_{bo}$ of 50%. To simplify our discussion on the ^{13}C content in this section, we will use the expected $^{13}\delta_{bo}$ for two alternative $^{13}\delta$ values of the soil CO_2: $-25‰$ for temperate climatic vegetations, and $-13 \pm 2‰$ for semi-arid regions (Section 5.1.4.3). The $^{13}\delta$ value of soil limestone is assumed to be $+1‰$. The $^{13}\delta_{bo}$ value matching these conditions will be about -12 or $-6‰$, respectively. During the last few years many examples have been given of ^{14}a values for groundwater samples that exceed 100%. This is caused by the fact that since the early 1960s recent vegetation contains nuclear bomb ^{14}C.

9.4.1.2   *Atmospheric CO_2 in rain water*

The given dissolution process might also be caused by atmospheric CO_2 dissolved in the rain water, instead of the organic soil CO_2. The concentration of atmospheric CO_2 ($\sim 0.03\%$) is in most cases negligibly small, as compared to that of CO_2 in the soil air (in the order of several per cent).

The relatively small direct contribution of atmospheric CO_2 to carbonate dissolution is confirmed by the slow response of the groundwater ^{14}C content to the sudden rise in atmospheric ^{14}C level around 1963.

However, in areas of little or no vegetation, where the groundwater tends to be very soft, the contribution of atmospheric CO_2 must be seriously considered. Because this has a $^{13}\delta$ value of about $-8‰$, the bicarbonate will then show relatively large $^{13}\delta$ values around $-3.5‰$ ($= \frac{1}{2}[-8 + 1]$).

9.4.1.3   *Dissolution of silicate rock*

Especially in regions free of carbonates, soil CO_2 also participates in the dissolution of silicate rock according to reactions such as:

$$CaAl_2(SiO_4)_2 + 2CO_2 + H_2O \rightarrow Ca^{2+} + Al_2O_3 + 2SiO_2 + 2HCO_3^- \qquad (9.11)$$

The $^{14}a_{bo}$ and $^{13}\delta_{bo}$ values in this case are 100% and $-25‰$. The silicate weathering is considered to be a very slow process. Therefore, the actual contribution of this source of CO_2 to the groundwater carbon inventory is usually negligible.

9.4.1.4   *Action of humic acids*

Part of the H^+ ions in acid soil water can originate from humic acids:

$$2CaCO_3 + 2H(Hum) \rightarrow Ca(Hum) + Ca^{2+} + 2HCO_3^- \qquad (9.12)$$

giving $^{14}a_{bo} = 0\%$ and $^{13}\delta_{bo} = +1‰$. Generally the humic acid content of soil water is very small. However, in areas with peat bogs or peat layers this process forms a considerable contribution to the inorganic carbon content (C_T) of the groundwater.

The next processes refer to subsequent alterations of the ^{14}C and ^{13}C content of the groundwater due to generation of CO_2 in the saturated zone.

Concerning this, a point of general interest needs to be made. Contrary to the specific ^{14}C activity ($^{14}C/C$) in the water, the ratio is not affected by the injection of *dead carbon* – that is, without ^{14}C – into the aquifer. Therefore, in some cases the consideration of $^{14}C/H_2O$ would lead to better conclusions than the use of $^{14}C/C$ ($= ^{14}A$).

9.4.1.5   *Decomposition of organic matter in the saturated zone*
During a later stage of groundwater movement in an aquifer, carbon can be added through oxidation of organic material (peat):

$$6n\ O_2 + (C_6H_{10}O_5)_n + n\ H_2O \rightarrow 6n\ H^+ + 6n\ HCO_3^- \qquad (9.13a)$$

followed by:

$$m\ H^+ + m\ CaCO_3 \rightarrow m\ Ca^{2+} + m\ HCO_3^- \qquad (9.13b)$$

$^{14}a_{bo}$ depends on the age of the organic compound. Moreover, during this addition of bicarbonate to the water, the water itself might already be of a considerable age. If $m = 6n$, $^{13}\delta_{bo}$ will generally be about $-12‰$, for $m = 0$, $^{13}\delta_{bo} = -25‰$.

9.4.1.6   *Anaerobic decomposition of organic matter*
Anaerobic decomposition of organic matter (peat) generates CH_4 as well as CO_2. The fractionation involved causes the methane $^{13}\delta$ to be very low ($^{13}\delta < -60‰$), while ^{13}C is strongly enriched in the carbon dioxide: $^{13}\delta > 0‰$. This process is not seldom observed. In the provinces of Groningen and Friesland, the Netherlands, $^{13}\delta$ values up to $+10‰$ were found in relatively deep groundwater (50–120 m), while the water also contained hydrocarbons.

9.4.1.7   *Sulphate reduction*
In the absence of free oxygen, sulphate reduction might occur in the aquifers:

$$SO_4{}^{2-} + CH_4 \rightarrow HS^- + H_2O + HCO_3^- \qquad (9.14)$$

where $^{14}a_{bo}$ completely depends on the age of the methane. In this case $^{13}\delta_{bo}$ can have strongly deviating values (down to very negative), because CH_4 generated by bacterial decomposition of organic matter already has a low ^{13}C content (down to $-90‰$).

9.4.1.8   *Ca–Na exchange*
If the Ca^{2+} concentration in the aquifer decreased due to Ca–Na exchange additional carbonate can be dissolved, thus maintaining saturation with calcite. The $^{13}\delta$ and ^{14}a values will increase and decrease, respectively, while the carbon content of the water increases.

9.4.1.9   *CO_2 from volcanic activity*
Another source of CO_2 can be volcanic activity. Magmatic CO_2 is generally derived from limestone and consequently contains no ^{14}C. The original ^{14}C activity of the water may therefore be seriously affected by the additional solution of limestone, as well as by isotopic exchange. There is no simple method by which the effect on the $^{13}\delta$ value can be predicted. In these cases it might be more useful to use the $^{14}C/H_2O$ ratios instead of $^{14}C/C$.

9.4.1.10   *Exchange with the soil carbonate*

Another process which might affect the ^{14}C (and ^{13}C) content of groundwater is exchange with the soil carbonate or limestone in the unsaturated zone, or in the aquifer, whereby the ^{14}C content of the dissolved bicarbonate would decrease. This problem has been discussed by several authors. We should clearly distinguish between two different processes which can occur without affecting the average dissolved carbon content of the water.

9.4.1.10.1   Isotopic exchange

Isotopic exchange, involving a solid state diffusion of CO_3^{2-} ions through the calcite crystal, is considered to be extremely slow in massive carbonates.

In soil carbonates, however, a much larger surface area is exposed to exchange. The fact that a thin surface layer of the carbonate grains will soon have reached a state of isotopic equilibrium with the dissolved carbon does not mean that no further exchange takes place. The radioactive decay of ^{14}C inside the carbonate grains maintains a stationary concentration gradient which causes a continuing diffusion of ^{14}C towards the inside. This does not apply to the stable ^{13}C. Results from artificial spikes have shown that the influence of the exchange on the ^{14}C age of groundwater is limited, and depends on the grain size, the carbonate content of the aquifer relative to the dissolved carbonate concentration of the water and presumably on the pH of the water. In high-carbonate aquifers, especially at temperatures above normal, high ^{14}C ages are considered to be questionable. Data have been reported, however, contradicting this prediction; authors have observed no differences in $^{13}\delta$ as well as ^{14}a between a series of cold (15 °C) and hot (60 °C) springs in one system.

9.4.1.10.2   Dissolution–precipitation

Dissolution–precipitation of carbonates in the soil seems to be a frequently occurring process in regions subject to long periods of dryness and high surface temperatures. In regions of temperate climate this process can be induced by seasonal temperature changes. The precipitation is mostly confined to a small soil zone with relatively low carbonate content. By comparing the masses of solid calcium carbonate involved and the dissolved bicarbonate seeping down annually, we feel that this process will generally be of minor importance. That part of the soil carbonate dissolved by the infiltrating water might contain some ^{14}C, and that the $^{13}\delta$ value is somewhat below the normal marine value, is taken care of in the equations of Section 9.4.4.

The given conclusions do not generally apply to regions with hydrothermal activity; at high temperatures (above 100 °C) the isotopic exchange might be a process of considerable importance.

Considering the number of possible carbon sources, the situation seems rather discouraging. However, the chemical composition of groundwaters indicates that in most cases the dominating process is that described by Eq. 9.10, although the possibility of interfering factors should be kept in mind. Therefore, before a ^{14}C analysis of groundwater is performed it is essential that a chemical analysis of the sample is available.

9.4.2   The CO_2–$CaCO_3$ concept

The basis of the method for dating groundwater by means of ^{14}C is illustrated in the schematic diagram for the origin and pathway of ^{14}C in Fig. 9.5. In order to clarify the origin of ^{14}C in a groundwater sample more specifically, we have to rewrite Eq. 9.10 in more complete form:

$$(a + \tfrac{1}{2}b)CO_2 + \tfrac{1}{2}bCaCO_3 + H_2O \rightarrow \tfrac{1}{2}bCa^{2+} + bHCO_3^- + aCO_2 \qquad (9.15)$$

In the soil $CaCO_3$ is dissolved by an equal molar amount of biogenic CO_2, whereas an additional amount of CO_2 is required to stabilise the solution chemically. For the ^{14}C analysis of groundwater, all CO_2 is extracted from the solution after acidification.

As mentioned in Section 9.4.1.1, the dissolved bicarbonate is expected to have a ^{14}C content of 50%, while an additional amount of CO_2 (with $^{14}a_{go} = 100\%$) shifts the ^{14}C content to a somewhat higher value. The early experience was that in many cases the recent groundwater values are up to 85%, although generally no statement was made about the original ^{14}C content. In some cases definite conclusions can be drawn because of the fact that the ^{14}C content of the groundwater sample is very low ($<5\%$) or very high ($>100\%$).

However, conclusions about groundwater flow directions and velocities can be deduced from *age differences* (= *relative ages*) within the aquifer. The ages are calculated after defining the upper, supposedly recent sample as being of zero age:

$$T = -8270 \ln \frac{^{14}a_m}{^{14}a_{recent}} \qquad (9.16)$$

This is certainly the most careful way of handling groundwater ^{14}C data. In fact, many hydrogeologists feel that ^{14}C will never provide us with more accurate groundwater ages.

On the other hand, situations occur where knowing the *absolute age* of even a single groundwater sample is relevant. For example, the isotope hydrologist is occasionally asked whether new boreholes for drinking water supplies might have a chance of carrying recent pollution; we will offer an answer to such issues in the next section.

In our discussion of the CO_2–$CaCO_3$ concept we have to distinguish between the closed system and the open system with respect to CO_2 (cf. Appendix III). In the first the carbonate is dissolved by infiltrating water which has initially been in contact with a CO_2 reservoir (the soil CO_2). In the open system the carbonate is dissolved by water in continuing contact with the CO_2 reservoir at a fixed partial CO_2 pressure. In nature, the carbonate dissolution will often take place under mixed conditions.

9.4.3   The closed dissolution system

In a closed system with respect to CO_2 the water reaches equilibrium with the gaseous CO_2.

9.4.3.1   *The chemical dilution correction*
Attempts have been made to correct the measured ^{14}C content or, equivalently, to calculate the original ^{14}C content of the water ($^{14}a_\Sigma$) from other data. This subject has been briefly discussed in Section 6.1.4.2.

Assuming Eq. 9.15 to be the only process taking place during groundwater recharge, the ^{14}C content of the total DIC with concentration Σ will be:

$$\Sigma^{14}a_\Sigma = \tfrac{1}{2}b^{14}a_1 + (a + \tfrac{1}{2}b)^{14}a_{ao} \qquad (9.17)$$

Here the total DIC is considered to comprise two fractions: dissolved bicarbonate (b) and dissolved CO_2 (a), while the dissolved CO_3^{2-} is neglected: $\Sigma = a + b$. Assuming the ^{14}C activity of the soil carbonates ($^{14}a_1$) to be equal to zero – an assumption which might not be valid (Section 9.4.1.9 and Figs 6.4 and 6.6) – the ratio between the original and the natural ^{14}C content is:

$$\frac{^{14}a_\Sigma}{^{14}a_{ao}} = \frac{a + 0.5b}{a + b} = \frac{\Sigma - 0.5b}{\Sigma} \tag{9.18}$$

The resulting age is:

$$T = -8270 \ln \frac{^{14}a_m}{^{14}a_\Sigma} = -8270 \ln \frac{^{14}a_m/^{14}a_{ao}}{^{14}a_\Sigma/^{14}a_{ao}} = T_{conv} + 8270 \ln F \tag{9.19}$$

The factor $^{14}a_\Sigma/^{14}a_o$ (sometimes given as F, Q or q in the literature) is often referred to as the *dilution factor* – incorrectly, because with high dilution (of natural ^{14}C-active carbon by inactive carbon from limestone) the factor is low. The use of this factor implies that from the conventional age (first term) a correction term is subtracted which follows from a measurement of the dissolved CO_2 and bicarbonate concentrations in the sample. In practice, b can be simply determined by acid titration in the field or in the laboratory (the CO_3^{2-} concentration is negligibly small in most cases); a + b results from an alkali titration. As an alternative a can be calculated from a pH measurement (preferentially in the field), using known values for the first apparent dissociation constant of carbonic acid (Appendix III).

Although cases have been reported in which the chemical dilution correction appears to offer reasonable results from a hydrogeologic point of view, some fundamental objections can be raised. The addition of carbon from other sources to the groundwater mass is being ignored, as well as later changes in the aquifer pH. Additional H^+ from, for example, humic acid would decrease b in Eq. 9.18 without affecting the ^{14}C content.

The fact that part of the dissolved carbon in groundwater occasionally originates from carbonates with low or zero ^{14}C content is referred to as the *hard water effect*. By this is meant the age difference between conventional dating ($^{14}a_o = 100\%$) and the corrected ages ($50\% < {}^{14}a_\Sigma < 100\%$). Assuming $^{14}a_\Sigma$ to be 85%, the hard water effect is 1350 years. The name of the effect is somewhat confusing, because theoretically the effect is not necessarily larger, the harder the water.

9.4.3.2 *The isotopic dilution correction*

An alternative and effectively equivalent procedure to correct ^{14}C ages from groundwater has now been introduced. The principle is that the chemical origin of the total dissolved inorganic carbon determines the stable carbon isotopic composition, as it does the ^{14}C content.

From Eq. 9.15 we derive:

$$\Sigma^{13}\delta_\Sigma = \frac{1}{2}b^{13}\delta_l + \left(a + \frac{1}{2}b\right) {}^{13}\delta_{ao}$$

where $^{13}\delta_l$, $^{13}\delta_{ao}$ and $^{13}\delta_\Sigma$ refer to the limestone, the biogenic CO_2 and the total carbon content, respectively.

This equation hence gives:

$$\frac{^{13}\delta_\Sigma - {}^{13}\delta_l}{^{13}\delta_{ao} - {}^{13}\delta_l} = \frac{a + 0.5b}{a + b} \tag{9.20}$$

Comparing Eqs 9.18 and 9.20 shows that the 'dilution factor' can also be written in terms of $^{13}\delta$ values:

$$\frac{^{14}a_{\Sigma}}{^{14}a_{ao}} = \frac{^{13}\delta_{\Sigma} - ^{13}\delta_{l}}{^{13}\delta_{ao} - ^{13}\delta_{l}} \qquad (9.21)$$

The choice of proper value for the ^{13}C content of soil carbonate ($^{13}\delta_{l}$) is not very critical. Marine limestone values normally range from 0 to $+2‰$. However, due to dissolution – precipitation processes (Section 9.4.1.10.2) negative $^{13}\delta_{l}$ (to $-2‰$) values have been observed and used.

Larger natural variations are observed in $^{13}\delta_{go}$, depending on the type of vegetation in the recharge area. According to the photosynthetic process of CO_2 assimilation, three groups of terrestrial plant material that contribute to the soil CO_2 generation can be distinguished: the Calvin group, showing $^{13}\delta_{go}$ of $-27 \pm 5‰$, the Hatch-Slack group having $^{13}\delta_{go}$ values of $-12 \pm 2‰$, and the CAM type showing a large spread around $-17‰$ (Section 5.1.4.3).

In moderate climates with mixed vegetation, plant decay in the soil generates CO_2 with an average $^{13}\delta$ of $-25‰$; direct measurements have shown $^{13}\delta_{go}$ values of about $-25‰$. In semi-arid regions, however, where plants often obey the Hatch-Slack cycle, much higher $^{13}\delta_{go}$ (and consequently less negative $^{13}\delta_{b}$) values are observed in the range -12 to $-14‰$.

Because of the process of diffusion of CO_2 from the root zone to the air, isotope fractionation occurs. This may cause the CO_2 present in the soil to have a $^{13}\delta_{go}$ value that is a few ‰ less negative than $-25‰$. In regions of temperate climate $^{13}\delta_{go}$ values of about -22 to $-25‰$ are observed. In semi-arid regions great caution is necessary when applying the isotopic dilution correction for obtaining true groundwater ages. In this respect, the chemical dilution correction will cause fewer uncertainties.

In cases where the actual $^{13}\delta$ values are reasonably well known, the closed-system concept should provide equal dilution factors, irrespective of whether they are chemically or isotopically derived.

9.4.4   The open dissolution system

In an open system with respect to CO_2 calcite dissolution occurs while the solution remains in contact with the surrounding gaseous CO_2.

9.4.4.1   *Isotopic exchange in the unsaturated zone*

The closed dissolution system presumably applies to (semi-arid) regions with little generation of soil CO_2 and where recharge occurs during short, heavy rainfall, thus washing the biogenic CO_2 out of the soil zone. However, even under these circumstances the system might act as an open system. The fact is that CO_2 might also be generated in the zone of percolation by decay of organic matter washed down by the percolating water. Contrary to the common soil air CO_2, this CO_2 is referred to as ground air carbon dioxide. In most cases, at least in temperate climates, the inorganic carbon formed according to Eq. 9.10 can easily exchange isotopes with gaseous CO_2 in the unsaturated zone. The possible occurrence of this process changes the dissolved bicarbonate $^{13}\delta$, without affecting the chemical composition of the solution. Therefore, we have to be cautious when simply applying the isotopic dilution correction (Section 9.4.3.2).

The isotopic exchange is assumed to explain the fact that a value of $85 \pm 5\%$ for the ^{14}C content of recent groundwater frequently occurs. This conclusion was drawn from a histogram of many ^{14}C data. However, recent waters are also found with lower ^{14}a values (65 ± 5%). The combined action of the dissolution as well as the exchange process, thus determining the chemical and isotopic composition of the groundwater, should be treated rigorously.

9.4.4.2 The dissolution-exchange model

The *DE-model* is, in principle, and according to the original ideas relating to ^{14}C, used as a tool for dating groundwater: the ^{13}C and ^{14}C content of the groundwater bicarbonate resulting from dissolution is calculated from mass balances, based on the observed chemical composition of the sample. The (partial) isotopic exchange between the dissolved bicarbonate and CO_2 in the unsaturated zone is subsequently deduced from the shift in $^{13}\delta(HCO_3^-)$.

Suffice it to state here the conditions, and give a brief outline with the resulting equations. The DE-model only applies in cases where:

- The reaction of Eq. 9.15 is the only source of dissolved carbon.
- No exchange occurs between the solution and the solid soil carbonate.
- The pH is solely determined by the dissolution process.
- The water sample does not comprise a mixture of waters of different age.

A schematic diagram of the ^{14}C and ^{13}C concentrations according to the DE-model is presented in Fig. 9.6. Essential steps in the formation of the dissolved carbon are as follows:

Step 1   The soil carbonate reacts with CO_2 dissolved in the infiltrating rain water (Eq. 9.10). Depending on the actual conditions, ^{14}a and $^{13}\delta$ values for the gaseous CO_2 (go) or for the CO_2 dissolved in isotopic equilibrium (ao) should be taken; $\varepsilon_{a/g} \approx -1\%o$.

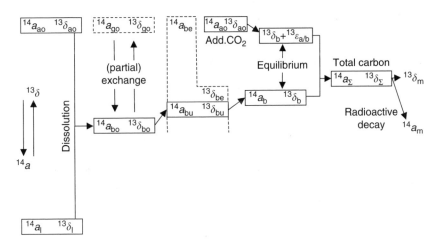

Figure 9.6.   Diagram explaining the DE-model for determining the original ^{14}C content of a groundwater sample. The evolution of DIC in groundwater is represented by three steps: (1) Dissolution of $CaCO_3$ by dissolved CO_2; (2) (Partial) exchange between the dissolved HCO_3^- thus formed and gaseous CO_2 in the unsaturated zone; (3) Chemical stabilisation of the solution by an additional fraction of dissolved CO_2, and isotopic exchange between the dissolved HCO_3^- and CO_2 fractions.

The resulting equations for ^{14}C and ^{13}C are equivalent to those obtained earlier (Eq. 9.17):

$$^{14}a_{bo} = \frac{1}{2}\left(^{14}a_{ao} + {}^{14}a_1\right) \tag{9.22a}$$

and

$$^{13}\delta_{bo} = \frac{1}{2}\left(^{13}\delta_{ao} + {}^{13}\delta_1\right) \tag{9.22b}$$

where instead of $^{14}a_{ao}$ we can also write $^{14}a_{go}(= {}^{14}a_{ao} + 0.2\%)$ and, instead of $^{13}\delta_{ao}$, $^{13}\delta_{go}(= {}^{13}\delta_{ao} + 1\%o)$.

Concerning the actual ^{14}a and $^{13}\delta$ values for the soil carbonate, these might be altered from 'marine values' to mixed values because of the additional formation of secondary carbonate in the soil.

If the actual soil carbonate consists of a mixture of old marine limestone and recently formed calcium carbonate, an equation can be derived to relate the values of $^{14}a_{lo}$ and $^{13}\delta_{lo}$ with $^{14}a_1$ and $^{13}\delta_1$. For $^{14}a_{lo} = 0$, this is approximated by:

$$^{14}a_1 \cong {}^{14}a_{go}\frac{^{13}\delta_1 - {}^{13}\delta_{lo}}{^{13}\delta_{go} - {}^{13}\delta_{lo} - \varepsilon_{g/s}} \tag{9.23}$$

where $\varepsilon_{g/s} \approx -10\%o$, slightly depending on temperature (cf. Table 5.2). If the secondary carbonate, however, is of considerable age the calculated value for $^{14}a_\Sigma$ will become too high, and consequently also the resulting groundwater age.

Step 2   The bicarbonate fraction exchanges isotopically with the gaseous CO_2 in the unsaturated zone, causing a parallel shift in ^{14}C and ^{13}C concentrations towards, but not necessarily reaching, an isotopic equilibrium situation (equal to calcite saturation in an open system):

$$\frac{^{14}a_{bu} - {}^{14}a_{bo}}{^{14}a_{be} - {}^{14}a_{bo}} = \frac{^{13}\delta_{bu} - {}^{13}\delta_{bo}}{^{13}\delta_{be} - {}^{13}\delta_{bo}} \tag{9.24}$$

where the complete exchange is represented by:

$$^{14}a_{be} = {}^{14}a_{go}(1 - 2{}^{13}\varepsilon_{g/b}) \tag{9.25a}$$

and

$$^{13}\delta_{be} = {}^{13}\delta_{go} - {}^{13}\varepsilon_{g/b}(1 + {}^{13}\delta_{go}) \approx {}^{13}\delta_{go} - {}^{13}\varepsilon_{g/b} \tag{9.25b}$$

Step 3   The solution finally contains a certain fraction of dissolved CO_2 in chemical equilibrium with the bicarbonate (a). This effects an isotopic shift from $^{13}\delta_{bu}$ to $^{13}\delta_b$. Furthermore, the total carbon content is extracted from the sample for the isotope analysis:

$$b^{14}a_{bu} + a^{14}a_{ao} = \Sigma^{14}a_\Sigma \tag{9.26a}$$

and

$$\begin{aligned} b^{13}\delta_{bu} + a^{13}a_{ao} &= b^{13}\delta_b + a^{13}\delta_a + c^{13}\delta_c \\ &\approx b^{13}\delta_b + a(^{13}\delta_b + {}^{13}\varepsilon_{a/b}) \\ &= \Sigma^{13}\delta_b + a^{13}\varepsilon_{a/b} \end{aligned} \tag{9.26b}$$

The resulting initial ^{14}C activity of the groundwater is then:

$$^{14}a_\Sigma = \{(a + 0.5b)/\Sigma\}^{14}a_{ao} + (0.5b/\Sigma)^{14}a_l + \Big\{^{14}a_{go}(1 - 2^{13}\varepsilon_{g/b})$$

$$-0.5(^{14}a_{ao} + {}^{14}a_l)\Big\} \times \frac{^{13}\delta_\Sigma - \{(a + 0.5b)/\Sigma\}^{13}\delta_{ao} + (0.5b/\Sigma)^{13}\delta_l}{^{13}\delta_{go} - {}^{13}\varepsilon_{g/b}(1 + {}^{13}\delta_{go}) - 0.5(^{13}\delta_{ao} + {}^{13}\delta_l)}$$

$$(9.27)$$

The first two terms refer to the closed-system dissolution process, and the third represents isotopic exchange with the soil CO_2. As mentioned before, $^{14}a_{go}$ and $^{13}\delta_{go}$ may be taken instead of $^{14}a_{ao}$ and $^{13}\delta_{ao}$ ($^{13}\delta_{ao}$ is about 1‰ more negative than $^{13}\delta_{go}$). It should be noted, however, that from a theoretical point of view the $^{14}a_{go}$ value depends on the $^{13}\delta_{go}$ value chosen, since $^{14}a_{go} = 100\%$ only if $^{13}\delta_{go}$ is $-25‰$ (by definition). For $^{13}\delta_{go} = -24‰$ the $^{14}a_{go}$ has to be raised to 100.2%, and at $^{13}\delta_{go} = -12‰$, $^{14}a_{go}$ is automatically 102.6%. In practical applications this is only a minor adjustment.

The minimum chemical information required is the sample pH. The carbonic acid chemistry is treated in Appendix III. Values for $^{13}\varepsilon$ have been given in Table 5.2.

9.4.4.3   *The isotopic parameters*

If the groundwater sample is expected to be at least several tens of years old, $^{14}a_{ao}$ ($\approx {}^{14}a_{go}$) has to be taken as 100%. In moderate climates $^{13}\delta_{go} \approx -24 \pm 2‰$, and in semi-arid regions $-12‰$ at a maximum ($-15‰$ is more likely). Values for $^{13}\delta_l$ generally vary between $+2(^{13}\delta_{lo} = 1 \pm 1‰)$ and $-2‰$; in the latter case $^{14}a_l$ is expected to be close to 20% instead of $^{14}a_{lo} = 0\%$. For recently infiltrated water $^{14}a_{go}$ can exceed 100%, depending on the vegetation.

The value for $^{14}a_\Sigma$, calculated from Eq. 9.27 does not vary for the three DE steps. It appears to make no difference whether:

A   the final bicarbonate resulted from first dissolution and subsequent exchange with CO_2,
B   or exchange took place during the carbonate dissolution process,
C   or the sample consists of a mixture of fractions which have separately undergone the processes A and B, assuming that all fractions are of recent age.

The value of the DE-model is, primarily, that it presents a framework for judging measured ^{14}C values. Even where it is not possible to deduce reliable absolute ages, it presents the possibility of putting limits to the water age.

9.4.5   Isotopic changes in the aquifer

Once the chemical and isotopic composition of the potential groundwater has been determined in the unsaturated zone, further composition changes might occur below the phreatic surface in the direction of flow. In order to describe these changes specific models exist, based on the presence of additional carbon sources and sinks in the aquifer. Discussion of these is beyond the scope of this textbook.

9.5   RELATIVE ^{14}C AGES OF GROUNDWATER

In the foregoing sections we have dealt primarily with attempts to obtain absolute groundwater ages. Although these are important to know in several situations, the isotope hydrologist is frequently confronted with the determination of groundwater flow direction and velocity. For solving these problems it is generally sufficient to know the relative ages of a series of water samples. Assuming that the water mass has a recharge area with a homogeneous vegetation, and that this plant cover did not change over the period of time covered by ^{14}C dates, $^{14}a_\Sigma$ may be taken equal for all samples. The relative age of sample $k+1$ relative to sample k is then simply:

$$\Delta T = -8270 \ln(^{14}a_{k+1}/^{14}a_k) \qquad (9.28)$$

As mentioned before (Section 9.4.2), this has been applied to several cases.

If the additional carbon is assumed to be of zero $^{14}a_1$ (limestone), $^{14}a_{k+1}$ has to be corrected by a factor Σ_{k+1}/Σ_k so that the relative age becomes:

$$\Delta T = -8270 \ln \frac{^{14}a_{k+1}}{^{14}a_k} \frac{\Sigma_{k+1}}{\Sigma_k} \qquad (9.29)$$

If no correction is applied for additional carbonate dissolution, too large age differences result; the flow velocities will therefore be overestimated. The flow direction of the water, however, can be established beyond doubt.

9.6   DATING GROUNDWATER WITH DOC

Attempts have been made to determine the age of groundwater, that is, the time elapsed since infiltration, by means of dating the DOC content of the water sample. Based on their solubility in specific acid solutions, the soluble and thus mobile organic compounds are subdivided into *fulvic acids* (soluble at low pH) and *humic acids*. These organic molecules, originating from the decomposition of organic matter, are relatively resistent to further degradation.

Here we are also confronted with the problem of assigning $^{14}a^i$ values to the recently infiltrated water. The fulvic fraction originates with certainty in the soil zone and is thus able to produce reliable ^{14}C ages, based on an $^{14}a^i$ value in the range of $85 \pm 10\%$. This range of values depends on the average age of the soil organic matter in the recharge zone, which can be hundreds of years.

9.7   RELATION BETWEEN ^{14}C AND ^{3}H IN GROUNDWATER

It is essential that a groundwater sample intended for ^{14}C measurement is also used to determine the ^{3}H content. If the sample shows a significant ^{3}H activity (>1 TU), the water – or at least one of its components if it consists of a mixture of old and young water – is subrecent (<50 years old). The ^{14}C and ^{3}H data, when combined with hydrogeological evidence, often present the possibility for estimating the origin of the groundwater, and for deciding about the relevance of a ^{14}C age and about the validity of a specific ^{14}C correction method.

> Again we must emphasise that it is unrealistic to try to find a generally valid procedure for simply 'measuring' the absolute age of a single groundwater sample using ^{14}C and ^{3}H.

Although many parameters are unknown, such as the mixing ratio of old and young water, the original ^3H content of the corresponding precipitation, the ^{14}C content for the source of the organic CO_2, etc., in combination ^{14}a and ^3A data can put limits to the age of the sample (Figs 9.8 and 9.10).

Concerning ^{14}C, the reality is that water infiltrates through a vegetation layer containing carbon from more than one year's growth (in Fig. 9.8 20 and 50 years, respectively). This causes a 'broadening' of the ^{14}C input function rather than the most simple original annual ^{14}C distribution in atmospheric CO_2 (Fig. 9.7). The more likely ^{14}C input shown in Fig. 9.8 is obtained by calculating ^{14}C using a 'moving average' procedure over 20 and 50 years.

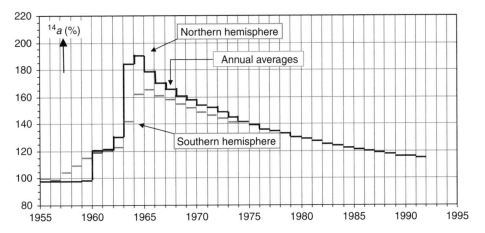

Figure 9.7. ^{14}C content of atmospheric CO_2 averaged over single years for the Northern and Southern hemispheres, respectively (cf. Fig. 6.3).

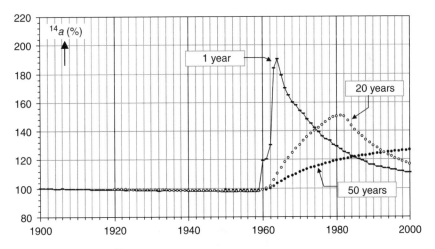

Figure 9.8.   Pattern of the ^{14}C distribution in atmospheric CO_2 for yearly averages (cf. Figs 6.3 and 9.7), and for the mean values over 20 and 50 years, calculated by a moving average procedure (each data point represents the average value during the period of the year indicated and the previous 20 or 50 years).

The effect is obvious: differences occur in the order of tens of per cent, equivalent to hundreds of years in age.

For ^{3}H the situation is slightly different, although the effect is analogous. Realistically, we have to accept lower values than the peak values of Fig. 9.9, and for two reasons:

1. The ^{3}H content of water has considerably decreased by radioactive decay since it was precipitated.
2. The main season for groundwater recharge is generally the winter period, during which the ^{3}H content is relatively low (Fig. 9.9).

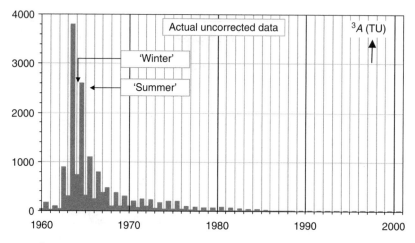

Figure 9.9. ^{3}H content of precipitation representative for the Northern hemisphere. The 'summer values' (April–September) are significantly higher than the 'winter values'. This is important when observing and estimating the actual infiltration of precipitation (cf. Fig. 6.8).

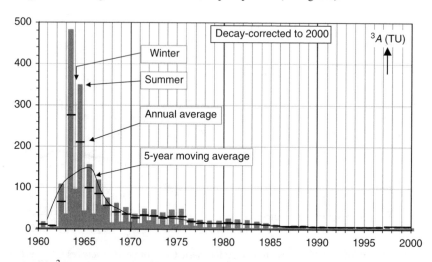

Figure 9.10. ^{3}H content of precipitation during the winter (October–March) and summer periods (April–September) for single years and for whole years (data represent the Northern hemisphere and are corrected for radioactive decay to the year 2000) (cf. Fig. 9.9).

There is still another reason why ^3H in groundwater may be lower than expected; that is the broadening effect caused by dispersion of the aqueous constituents amongst which $^3H^1HO$.

The drawn curve in Fig. 9.10 is the result of a 5-year moving average procedure, representing the effect of dispersion of aqueous constituents as well as tritiated water molecules in the infiltrating water. These data are illustrative, rather than for direct use in actual case studies; similar procedures then have to be followed using ^3H data from the IAEA/WMO network, applicable to the region under study (available online at: http://isohis.iaea.org).

APPENDIX I

Water sampling and laboratory treatment

This chapter on field and specifically on laboratory practice serves to give the hydrologist as well as the beginning laboratory worker an impression of the experimental efforts involved in applying isotope hydrology. A few parts have been taken – and slightly adjusted – from the IAEA Guidelines/Manual for Operation of an Isotope Hydrology Laboratory. Detailed descriptions of the field practice are given by Clark and Fritz (1997) (see Recommended literature). We have limited the amount of detail on the individual analytical methods, so that the contents of (parts of) this chapter are valid for any isotope laboratory.

I.1 WATER SAMPLING AND STORAGE

In a successful field sampling programme a volume of water is to be collected and brought or sent to the isotope laboratory that is representative for the water mass under investigation. The main concern during sampling, transport and storage is to avoid isotope fractionation through evaporation or diffusive loss of water vapour, and/or isotope exchange with the surroundings as well as with the bottle material. These effects can be minimised by using appropriate collection methods and bottles. A careful selection and testing of the methods and bottles needs to be made at an early stage. This section provides background information to help in this selection.

> The hydrologist should, however, follow the specific instructions given by the laboratory that is to analyse the samples.

I.1.1 Sampling bottles

The bottle and seal must be made of such a design and appropriate material that any loss by evaporation and diffusion to, or exchange of water with the surroundings is prevented. The isotope effect of evaporation can be significant: a 10% loss of sample results in an isotope enrichment of about 10‰ in ^{2}H and 2‰ in ^{18}O. The following conclusions are offered from experience:

- The most secure vessels for storage are glass bottles, allowing storage for at least a decade as long as the seal is not broken.
- High-density polyethylene is satisfactory for collection and storage during a few months (water and carbon dioxide easily diffuse through low-density plastics).
- Bottles with small necks are the most satisfactory.

- Caps with positive seals (plastic inserts, neoprene, etc.) are required.
- The volume of bottles should normally be 50 mL for the combination of ^{2}H and ^{18}O analyses, 50 L for conventional ^{14}C dating, 250 mL for AMS ^{14}C dating (Appendix II), and 500 mL for low-level ^{3}H analysis. If the storage is expected to be longer than several months, it is preferable to collect a larger volume and store the samples in glass bottles (the relative influence of evaporation is than minimised). The dissolved carbon content of large-volume ^{14}C samples are generally treated in the field (see Section I.1.2).

Details on sample quantities are given in Table II.2.

I.1.2   General field practice

The following points emphasise good field practice, applicable to all types of water to be sampled (precipitation, surface water, groundwater):

- Always use a field note book to record observations, and record numerical information on sample collection sheets to be handed over to the laboratory.
- Determine geographical co-ordinates of sampling points using GPS or national co-ordinate systems, maps, aerial photographs, etc.
- Measure altitude, depth to the water table (groundwater), sample depth (depth into groundwater or surface water), well condition, condition of rain gauge, discharge (rivers, artesian wells), lake level, weather conditions, etc.
- It is very important to record other chemical and physical data to aid interpretation, such as water temperature, pH, alkalinity, conductivity and other possible chemical constituents.
- Fill sample bottles completely, provided the water has no chance of freezing during air transport (in that case the bottles should be two-thirds full).
- Mark all bottles individually with waterproof marker (project code, location, date, sample number, collector's name, type of analysis needed); the information has to be cross-referenced with the field notebook and the sample collection sheets.

Specific points of interest are further discussed below.
    In the following sections we discuss sampling of the various types of water.

I.1.3   Precipitation

The sampling strategy for rain and snow is in general dictated by the scientific aim of the programme. For example, the sampling interval may be monthly, weekly, daily or even each hour. In all cases it is necessary to record the amount of precipitation, so that weighted means of isotopic compositions can be calculated afterwards.

    The maximum information and least risk of evaporation result from short-term collection of precipitation. Samples should be stored individually in bottles. If only monthly mean compositions are required, daily samples can be combined in a larger storage bottle.

    For samples that are collected on a weekly or monthly basis, the isotopic composition of the sample is likely to be modified by evaporation. This can be minimised by a special construction of the rain gauge, or by placing a small amount of mineral oil (2 mm minimum) in the collection bottle which is floating on the water.

Special care needs to be taken when collecting snow; it should be allowed to melt slowly at ambient air temperature, avoiding evaporation.

I.1.4 Surface water

The collection methods for surface water generally pose few problems, particularly when it is collected in relatively small quantities. Special care has to be taken when collecting water for carbon isotope analysis; field measurement of temperature, pH and salinity (especially with marine and brackish water) are needed. The samples are to be stored in glass bottles, in the dark, preferably at low temperature, and should be treated with $I_2 + KI$ or $HgCl_2$. The poisoning recipes for this are:

> $I_2 + KI$ *solution*: prepared by dissolving 15 mg of I_2 and 30 mg of KI per mL of (ultra)pure water; water is poisoned by adding 0.5 mL of this solution per L of sample.

> $HgCl_2$ *solution*: prepared by dissolving 70 mg $HgCl_2$ per mL of (ultra)pure water; 3 mL of this solution is to be added per L of water sample.

River and stream samples should be taken from mid-stream or a flowing part of the stream. Standing water near stream banks should be avoided, as the water may not be well mixed and may show effects due to evaporation or pollution.

Samples from lakes and estuaries should be collected from both the surface and from depth. Together with other physical and chemical information, it may thus be possible to interpret the results in terms of the water-column structure.

Care should be taken when sampling near a confluence. The river samples may still show variable isotopic compositions for a considerable distance downstream of a confluence, due to incomplete mixing between the two different river waters. In large river systems this may amount to a distance of many tens of kilometres.

I.1.5 Unsaturated zone water

The isotopic profile of soil moisture can provide information on groundwater recharge. Samples are often taken from the soil as such. The most common methods to extract water from the soil profile in the laboratory are (i) vacuum distillation, (ii) freeze drying, (iii) squeezing and (iv) centrifuging.

I.1.6 Groundwater

For all groundwater sampling it is necessary to characterise as far as possible the hydrogeological situation (geophysical, geochemical, etc.) of the borehole.

Pumped boreholes and production wells present little problem, with the sample readily collected from the surface supply. This is specifically true for the large quantity samples needed for conventional ^{14}C analysis (see Appendix II). Sampling observation wells and especially very deep aquifers, may pose more problems. One should first be aware that standing water in such passive boreholes may not be representative for the water mass in the aquifer, because of evaporation or carbon dioxide exchange with the ambient air. It is usual to pump the borehole until it is purged by approximately twice the well volume, or until steady-state chemical conditions (pH, Eh) are obtained. While this is common practice, it should be borne in mind that in certain situations, the cone of depression resulting from pumping may draw in water from other sources.

It may be practically impossible to collect sufficient water from a deep well for routine ^{14}C analysis. In that case one has to turn to the AMS technique (Section II.3), for which 250 mL of sample is generally sufficient.

I.1.7   Geothermal water

Both steam and liquid phases should be collected from geothermal fields so that the ratio steam/water can be determined. This is relatively easy in production plants, but difficult in non-developed geothermal regions. The steam has to be condensed, taking care that this procedure operates quantitatively.

Particularly critical for thermal waters is the need to identify the source of the spring and to sample as close to the spring as possible.

I.2   LABORATORY TREATMENT OF WATER SAMPLES

The aim of this section is to offer the hydrologist some insight into the procedures used by the isotope laboratory to determine the isotopic compositions of the water samples that have been sent/brought to the laboratory. The purpose is not to provide the laboratory with detailed recipes for carrying out the analyses, for that the reader is referred to IAEA guidelines or specific laboratory operation manuals and literature. The procedures are discussed on the basis of the various isotopes.

Nearly all isotopic measurement techniques cannot be applied to water as such. Therefore, the underlying principle for all methods is that the water or the constituents to be analysed (such as dissolved carbonic acid) are converted into a chemical compound that can be used in the equipment. The specific requirement is that during this conversion no changes are introduced in the isotopic composition, or only changes that are accurately known.

I.2.1   The ^{18}O/^{16}O analysis of water

I.2.1.1   *Equilibration with CO$_2$ for mass spectrometric analysis*
Since water poses many problems in mass spectrometers – primarily because of its adhesion to metal, causing serious memory effects – the ^{18}O/^{16}O ratio of the water sample has to be transferred to a more suitable gas, such as carbon dioxide. The most common sample preparation method for $^{18}\delta$ in water is the CO$_2$–H$_2$O equilibration, in which about a millimole of CO$_2$ (10 to 20 mL) is brought into isotopic equilibrium with a few mL of sample water (generally at $25.0 \pm 0.2\ °C$):

$$H_2\,{}^{18}O + C^{16}O_2 \Leftrightarrow H_2\,{}^{16}O + C^{18}O^{16}O$$

An aliquot of the equilibrated CO$_2$ is then transferred to the mass spectrometer for ^{18}O/^{16}O measurement. Because the ^{18}O/^{16}O ratio in CO$_2$ and H$_2$O is different (see Section 5.2.2 for the value for $^{18}\alpha_{g/l}$) ^{18}O/^{16}O of the water sample changes; the $^{18}\delta$ value obtained should therefore be corrected for this isotopic shift. However, since the sample and international reference material (see Section 5.2.3) are treated equally (i.e. equal amounts of water and carbon dioxide and equal temperature), the correction is not relevant:

$$^{18}\delta(\text{water}) = \frac{^{18}R_{\text{H}_2\text{O}}}{^{18}R_{\text{VSMOW}}} - 1 = \frac{^{18}\alpha_{\text{g/l}}\,^{18}R_{\text{H}_2\text{O}}}{^{18}\alpha_{\text{g/l}}\,^{18}R_{\text{VSMOW}}} - 1$$

$$= \frac{^{18}R_{\text{CO}_2\text{ in equil. with H}_2\text{O}}}{^{18}R_{\text{CO}_2\text{ in equil. with VSMOW}}} - 1 = {}^{18}\delta(\text{equilibrated CO}_2) \qquad (\text{I.1})$$

In other words, the $^{18}\delta$ value for CO_2 equilibrated with the water sample is equal to the $^{18}\delta$ value for the water itself, where the $^{18}\delta$ values refer to the international reference VSMOW. We have to emphasise, however, that the δ values do have to be corrected for the water–CO_2 equilibration according to Eq. I.3, because of the additive term in Eq. I.2.

The following experimental details and concerns apply to this method:

- During the isotopic exchange of the oxygen atoms between CO_2 and H_2O the oxygen isotopic composition of the water changes; for this a correction has to be made. From an ^{18}O mass balance consideration we derive:

$$w_0\,^{18}R_{w_0} + c_0\,^{18}R_{c_0} = w^{18}R_w + c^{18}R_c = w\frac{^{18}R_c}{^{18}\alpha_{c/w}} + c^{18}R_c$$

where w_0 and c_0, and w and c are the mole amounts of oxygen atoms in the water and carbon dioxide before and after the equilibration, respectively, and the fractionation factor $^{18}\alpha_{c/w}$ is equal to $1 + {}^{18}\varepsilon_{g/l}$ as defined in Table 5.4 (commonly at 25 °C). Assuming that $w = w_0$ and $c = c_0$, and writing $\rho = c/w$, the corrected value for the equilibrated CO_2 is:

$$^{18}\alpha_{c/w}\,^{18}R_{w_0} = {}^{18}R_c(1 + {}^{18}\alpha_{c/w}\rho) - {}^{18}\alpha_{c/w}\rho\,^{18}R_{c_0} \qquad (\text{I.2})$$

In terms of δ values, this reduces to a corrected $^{18}\delta$ value for:

$$^{18}\delta_c' = (1 + {}^{18}\alpha_{c/w}\rho)\,^{18}\delta_c - {}^{18}\alpha_{c/w}\rho\,^{18}\delta_{c_0} \qquad (\text{I.3})$$

- Automated units can be procured with either an air bath or water bath. The former are easier to operate, but the latter show a better temperature stability and homogeneity throughout the bath.
- A variation in equilibration temperature of 1 °C causes an uncertainty in the $^{18}\delta_c$ value for 0.2‰. Therefore, the temperature variation of the bath over time should generally be much less than 0.5 °C, and preferably no more than 0.2 °C.
- The equilibration time depends on the amplitude and frequency of shaking and the amount of sample. Typical equilibration times are in the order of 4 to 8 hours.
- Two ways are being used to remove oxygen and other gases from the water sample prior to equilibration: (i) freezing the sample and removing the gases by pumping (repeated procedure) and (ii) pumping the water sample at ambient temperature through a capillary, thus minimising the evaporation (and consequently isotope fractionation) of the sample.
- The pH of the water needs to be in the range of 6 to 7 so that dissolved CO_2 and HCO_3^- are abundant; otherwise, a drop of concentrated H_3PO_4 may be added.

- In saline waters and brines, the equilibration technique determines the oxygen isotope activity rather than isotope concentration in the sample. Therefore, it is first necessary to determine the sample chemistry.

I.2.1.2   *Other methods*

Equilibrating CO_2 with the water sample is the most common method for measuring $^{18}O/^{16}O$ in water. There are other methods which are in an experimental stage or specifically applicable to certain problems. These will only briefly be mentioned.

1. Water can be converted quantitatively to O_2 by *fluorination* with bromiumpentafluoride:

$$6H_2O + 2BrF_5 \rightarrow 10HF + 2HBr + 3O_2$$

The oxygen released can be isotopically analysed directly, or can be treated with hot graphite to produce CO_2 for mass spectrometric analysis (O'Neil and Epstein, 1966). The advantage of this second method is that only 9 mg of water is needed. The disadvantage is that BrF_5 is an extremely dangerous liquid, and needs extended laboratory installations and precautions.
2. Water can be reacted in nickel tubes *with graphite* at high temperature to produce CO_2 (Edwards *et al.*, 1993).
3. The $^{18}O/^{16}O$ and $^{17}O/^{16}O$ ratios of water can be determined by mass spectrometer analysis of O_2 produced by *electrolysis*. The precision obtained is about 0.1‰; an advantage concerns the direct measurement of ^{17}O which is problematic in CO_2, since the ^{13}C containing molecule has the same mass and is much more abundant.
4. A promising new method is measuring isotope ratios ($^2H/^1H$, $^{17}O/^{16}O$, $^{18}O/^{16}O$) in water directly by *laser absorption spectroscopy*.

I.2.2   The $^2H/^1H$ analysis of water

I.2.2.1   *Reduction of water to H_2 for mass spectrometric analysis*

The introduction of water in the mass spectrometer has to be avoided. Therefore, the $^2H/^1H$ ratio is commonly transferred to a more suitable gas, *in casu* hydrogen. The production of H_2 gas is generally performed by *reducing* the water at several hundred degrees C with zinc or (depleted) uranium:

$$H_2O + Zn \rightarrow ZnO + H_2 \quad \text{at } 500\,^\circ C$$

$$\text{or} \quad 2H_2O + U \rightarrow UO_2 + 2H_2 \quad \text{at } 800\,^\circ C$$

The first method is generally carried out as a batch process in steel tubes. In the second method the evaporated sample moves through a furnace containing uranium turnings. Both options require about 10 μL of water. Although applying uranium has fewer experimental complications, the disadvantages are that it is poisonous, and more difficult to obtain and dispose of.

I.2.2.2   *Other methods*

Similar to the equilibration method for $^{18}O/^{16}O$ analysis, the $^2H/^1H$ ratio of a water sample can be transferred to H_2 by *isotopic exchange* at high temperature:

$$^2H^1HO + {}^1H_2 \Leftrightarrow {}^1H_2O + {}^2H^1H$$

As in chemical reactions, the equilibration technique is concerned with the activities rather than concentrations. Consequently, the chemistry of the water sample needs to be known. Other disadvantages are that the method needs a relatively large investment and that a very good temperature control is needed, as the isotope fractionation varies considerably with the exchange temperature (6‰ per °C). The advantage is that the procedure is faster than the other techniques.

A new method using *laser spectroscopy*, already mentioned under Section I.2.1.2, is still being developed but has been shown to be successful for $^2H/^1H$ analysis. The advantage of the method is that the water samples are measured as water vapour instead of as liquid water. There is therefore no need to apply the relatively troublesome decomposition of water into hydrogen gas.

I.2.3   The ^3H analysis of water

Several options are available for measuring the tritium content of water. The choice is determined by the (expected) ^3H content of the sample and the precision required. The different treatment procedures are:

1. Measuring the ^3H activity by *liquid scintillation spectrometry* (LSS).
2. Measuring the ^3H activity by *proportional gas counting* (PGC).
3. Measuring the amount of daughter ^3He mass spectrometrically, after storing the sample for a specific time in the laboratory.

Moreover, the possibility exists to increase the original ^3H content artificially (i.e. to enrich the sample) by a known amount, usually through *electrolysis*, prior to the ^3H measurement. Therefore, the additional options are:

(a) Treating the sample as it is collected.
(b) Treating the sample after artificial enrichment.

The range of choices is additionally broadened by the combination of (1), (2) or (3) with either (a) or (b). The choices to be made we discuss in Appendix II. Here we restrict ourselves to the different chemical/physical procedures.

Since the LSS technique is applied to water itself, the sample has to be mixed with a proper scintillation liquid (Appendix II). PGC requires a suitable gas, like stable isotopic analyses.

I.2.3.1   *Water purification*
The collected water often contains too many contaminants to allow direct LSS measurement as well as electrolytic enrichment. Coloured contaminants, in particular, hamper a proper liquid scintillation performance. *Purification* of water is carried out by *distillation*. In fact, most laboratories routinely subject all water samples to distillation, the requirement being that no isotopic change occurs during this treatment. Therefore, the distillation procedure has to be as quantitative as possible, that is, a yield close to 99%. The water sample should also be in minimal contact with the ambient air, since the vapour content may affect the ^3H content of the sample through exchange, an extremely fast process.

I.2.3.2 ^3H enrichment

Two methods are available for the artificial enrichment of ^3H in water:

1. Electrolysis of the water sample.
2. Thermal diffusion of H_2 prepared by reduction of water using, for example, zinc or magnesium. Very high enrichment factors can be obtained by this method, but the equipment and sample handling is complicated. It is not in common use in laboratories. The principle of *thermal diffusion* is that when a gas mixture is in contact with a cold wall as well as a hot wall, the high-mass (isotopically heavy) component tends to concentrate in the cold area and the low-mass (isotopically light) component in the hot zone. If this principle is applied to a vertical, double-wall tube structure in which the inner tube is heated and the outer wall is cooled, the heavy component concentrates at the cold wall and then, having a higher density (at low temperature) moves downward and concentrates at the lower end of the tubes. The hot, light component concentrates in the top volume. Consequently, a separation of the (isotopically) heavy and light molecules, in this case $^3H^1H$ and $^1H_2$ respectively, is obtained.

The first method is relatively simple, requires little sample handling or supervision during enrichment, and can be applied simultaneously to a series of samples including reference samples.

During electrolysis the purified sample (Section I.2.3.1) is provided with Na_2O or ^3H-free NaOH to increase the electrical conductivity, and is subsequently decomposed by an electric current:

$$2H_2O \rightarrow 2H_2 + O_2$$

The hydrogen isotope fractionation is remarkably large (contrary to the oxygen isotope fractionation): about 90% of the ^3H content of the original water sample remains in the water (the escaping H_2 gas is very much depleted in ^3H). This means that if a quantity of water is reduced by a factor of 10, the enrichment is about 9.

After electrolysis the enriched sample, now containing a large concentration of NaOH, is again distilled to remove the concentrated electrolyte prior to activity measurement.

Depending on the measurement technique to be applied, the amount of water needed is 5 to 20 mL. If the sample is to be enriched the desired quantity is 250 mL (see Table II.4).

I.2.3.3   *Preparation of gas for PGC of ^3H*

Several gases are suitable for proportional gas counting (Section II.3.1) of ^3H:

1. *Hydrogen*, prepared by

    $$H_2O + Zn \text{ or } Mg \rightarrow ZnO + H_2$$

    the prepared hydrogen gas has to be mixed with a proper counting gas such as argon.
2. *Ethane or propane*, prepared by the addition of hydrogen made as per (1) to unsaturated hydrocarbons, which are also suitable counting gases:

    $$C_2H_4 + H_2 \rightarrow C_2H_6 \quad \text{(with Pd catalyst)}$$
    $$C_3H_6 + H_2 \rightarrow C_3H_8 \quad \text{(with Pd catalyst)}$$

The additive reactions indicated under (2) cease at elevated temperatures in the presence of a suitable catalyst.

I.2.4   The ^{14}C analysis of dissolved inorganic carbon

I.2.4.1   *In the field*
The inorganic carbon content plays an important role in natural waters. As seen in the preceding chapters, the total DIC consists of varying concentrations of carbonic acid (H_2CO_3 or rather dissolved CO_2), dissolved bicarbonate (HCO_3^-) and dissolved carbonate (CO_3^{2-}). There are two methods in use for obtaining the DIC fraction in water and hence for measuring the ^{14}C content, or the ^{13}C/^{12}C ratio. Both, however, collect the total amount of DIC (to determine the separate $^{13}\delta$ values for the components we have to use the procedures discussed in Appendix III):

1. Addition of acid and extracting the gaseous CO_2 formed:

$$aCO_2(aq) + bHCO_3^- + cCO_3^{2-} + (b + 2c)H^+$$
$$\rightarrow (a + b + c)CO_2 + \ldots H_2O$$

Special, simple to operate equipment takes care that the escaping CO_2 is dissolved in a small amount of alkaline solution in a plastic bottle and shipped to the laboratory.

2. Precipitating DIC as $BaCO_3$ after adding NaOH solution:

$$aCO_2(aq) + bHCO_3^- + cCO_3^{2-} + (2a + b)OH^-$$
$$\rightarrow (a + b + c)CO_3^{2-} + \ldots H_2O$$
$$(a + b + c)CO_3^{2-} + BaCl_2 \rightarrow (a + b + c)BaCO_3 + \ldots Cl^-$$

Also here, the equipment (provided by the laboratory) is simple: the precipitate is collected in a small plastic bottle and shipped to the laboratory.

Both procedures may be carried out in the field, thus avoiding the need to transport a large quantity of water to the laboratory. Depending on the DIC content of the water, between 25 and 50 L is required for the conventional ^{14}C analysis. With the introduction of the AMS technique this problem no longer exists, since only a few mg of carbon are needed for this method. The same is valid for ^{13}C/^{12}C analysis. This means that 250 mL of water is generally sufficient for both carbon analyses. Additionally, no treatment is required in the field; the ^{13}C sample, however, needs careful storage (Section I.1.4).

I.2.4.2   *In the laboratory*
In principle, the initial procedures for treating the various samples received from the field are the same: treatment with acid (generally HCl solution) and purification of the escaping CO_2. Different methods are subsequently applied, depending on the ^{14}C measurement technique to be used. In chronological order these are:

1. *Activity measurement* in a PGC as CO_2, C_2H_2, C_2H_6 or CH_4

   (a) If CO_2 is used in the gas counter directly, it has to be extremely pure; this purification is the main effort of the sample treatment by the laboratory.
   (b) Preparing acetylene from CO_2 is performed in two steps:

   (i) $2CO_2 + 10Li \rightarrow 4Li_2O + Li_2C_2$ (700 °C)
   (ii) $Li_2C_2 + 4Li_2O + 6H_2O \rightarrow 10Li(OH) + C_2H_2$

(c) Preparing ethane from the acetylene prepared as under (b):

$$(i) \quad C_2H_2 + 2H_2 \rightarrow C_2H_6 \quad \text{(with Pd catalyst)}$$

(d) Preparing methane from CO_2:

$$(i) \quad CO_2 + 4H_2 \rightarrow 2H_2O + CH_4 \quad (300\,°C + \text{Ru catalyst})$$

2. *Activity measurement* in an LSS as C_6H_6. The acetylene gas prepared under 1(b) can be condensed to benzene:

$$3C_2H_2 \rightarrow C_6H_6 \quad \text{(with catalyst)}$$

The thus formed benzene is then mixed with a proper scintillation fluid (Appendix II).
3. *Concentration measurement* in an AMS as graphite or CO_2; in the presence of an iron catalyst CO_2 is reduced to graphite:

$$CO_2 + 2H_2 \rightarrow 2H_2O + C \quad (\text{at } 600\,°C \text{ with Fe})$$

The graphite is directly transferred to the accelerator target (see Appendix II).

I.2.5   The $^{13}C/^{12}C$ analysis of dissolved inorganic carbon

This procedure has been discussed in Section I.2.4.2, as it is similar – and in fact it runs parallel – to the ^{14}C procedure. The CO_2 extraction can be carried out by acidifying the water sample and collecting the escaping CO_2 during vacuum pumping. The sample can also be flushed with a clean gas, such as nitrogen or argon, and the escaping CO_2 likewise condensed in a cold trap.

One concern with these samples is that the water is likely to exchange with the atmosphere if it is not completely sealed from it. $^{13}C/^{12}C$ then ultimately changes to be in isotopic equilibrium with the atmospheric CO_2. At the same time the pH of the water changes to higher values. The sample should therefore be stored in a glass bottle and properly sealed.

A second concern is the growth of organic matter in the water sample. As this has a $^{13}\delta$ value much smaller than the DIC, the latter will increase. In order to avoid this process the sample needs to be stored at low temperature in the dark, preferably after poisoning (Section I.1.4).

APPENDIX II

Measuring techniques

The aim of this chapter is to give some insight to the various techniques for measuring isotope ratios and radioactivities. Potential readers are the hydrologists who apply the isotope techniques, their students assisting in the laboratory and laboratory technicians receiving a first glance of their work. The review is general in this respect, in that it covers the methodological field, although laboratories often do not have all techniques at their disposal. Since technical details vary between laboratories and manufacturers, we restrict ourselves to the underlying principles. A more detailed treatment of the analytical techniques is given in Volume I, chapter 11 of the UNESCO/IAEA series.

II.1 MASS SPECTROMETRY FOR STABLE ISOTOPES

II.1.1 Physical principle

A mass spectrometer is able to separate atoms, or rather ions, with different mass and measure their relative abundances. This principle is shown in Fig. II.1A where the gas containing different isotopes of an element (^{12}C and ^{13}C in CO_2, for example) is ionised in the ion source. The positive ions are accelerated by high voltage and subsequently enter a magnetic field perpendicular to the electric field (in Fig. II.1A drawn on the surface of the paper). The path of the ions now becomes circular because of the Lorentz force (Fig. II.1B). The circle radius depends on the ion mass: ions with heavier masses follow a larger circle. In this way the different isotopic ions become separated and can be collected. In these collectors the ions lose their electric charge, causing small electric currents that can be measured (Fig. II.2).

The magnitude of the Lorentz force F is:

$$\vec{F} = \vec{B} \times q\vec{v} \qquad (II.1)$$

This force is equal to the centripetal force required for keeping the particle in the circular path:

$$F = \frac{mv^2}{r} = Bqv \qquad (II.2)$$

The velocity of the particle is due to acceleration by the high voltage in the ion source (Fig. II.1B):

$$E_{kin} = \frac{1}{2}mv^2 = qV \qquad (II.3)$$

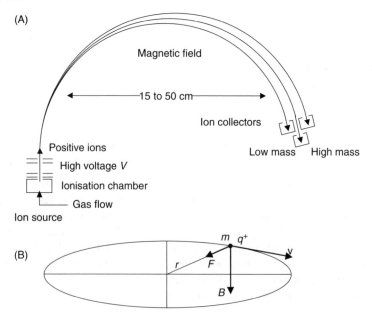

Figure II.1.   Schematic view of mass separation in a mass spectrometer. (A) The gas containing isotopic molecules enters the ion source and is ionised by electron bombardment. The positive ions are extracted from the ionisation chamber and are accelerated in the ion source by a high-voltage V. In a magnetic field perpendicular to the surface of the paper, the isotopically light and heavy ions are separated and collected by collectors. This results in electrical currents that can be measured very precisely. (B) Representation of the Lorentz force: a charged particle with positive electric charge q^+ and mass m moves in a flat surface with a speed v in a magnetic field B perpendicular to that surface; the resulting Lorentz force F is perpendicular to B and v and causes the particle to follow a circular path with radius r.

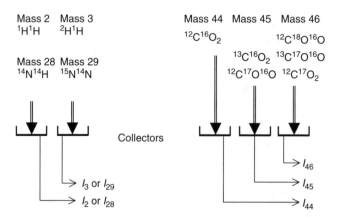

Figure II.2.   The various isotopic ions of hydrogen, nitrogen and carbon dioxide, respectively, are focused in the collectors (Faraday cups). These collect the electric charges (electrons are picked up by the positive ions) resulting in electric currents for the various isotopic molecules. From the ratios of these, the isotopic ratios of H_2 or N_2 (left) and CO_2 (right) can be calculated.

From Eqs II.2 and II.3 it follows that the radius of the circular path for the different isotopic ions is:

$$r = \frac{1}{B}\sqrt{\frac{2mV}{q}} = \left(\frac{\sqrt{2V/q}}{B}\right)\sqrt{m} \tag{II.4}$$

where q is equal to the electron charge $= 1.6 \times 10^{-19}$C, and B and V depend on the instrument settings. During a measurement the electric voltage and the magnetic field are kept constant, so that the radius is proportional to the square root of the ion mass: r (:) $\sqrt{m}$. Differentiation (dr/dm leading to the approximation $\Delta r/\Delta m$) results in:

$$\frac{\Delta r/r}{\Delta m/m} = \left(\frac{\sqrt{2V/q}}{B}\right) \quad \text{and} \quad m/\Delta m = \left(\frac{\sqrt{2V/q}}{B}\right) r/\Delta r \tag{II.5}$$

The latter is a measure of the *resolution of the mass spectrometer*: for $\Delta m = 1$, m is the *resolution* of the instrument if Δr is sufficiently large to allow a clear distinction between the peaks for masses m and $m + 1$ in the mass spectrum.

Isotope Ratio Mass Spectrometers (IRMS) have a typically low resolution (in the order of 100), contrary to the 'organic' mass spectrometers with resolutions of more than 10^4. The latter, on the other hand, have much lower sensitivity and are therefore not suitable for isotope ratio measurements.

II.1.2 Reporting stable isotope abundance ratios

The ion beams, comprising positive ions of the various isotopic molecules in the gas, are focused in the collectors. Here the electric charge is transferred to the metal collectors, resulting in electric currents (I) in the collector connections (Fig. II.2). From the ratios of these currents the isotopic ratios can be determined for the original gases. Each current is proportional (coefficients c) to the partial pressure of the respective component in the gas.

$$\begin{aligned}
I_2 &= c_2[^1H_2^+] & I_{44} &= c_{44}[^{12}C^{16}O_2^+] \\
I_3 &= c_3[^2H^1H^+ + {}^1H_3^+] & I_{45} &= c_{45}[^{13}C^{16}O_2^+ + {}^{12}C^{17}O^{16}O^+] \\
& & I_{46} &= c_{46}[^{12}C^{18}O^{16}O^+ + {}^{13}C^{17}O^{16}O^+ + {}^{12}C^{17}O_2^+]
\end{aligned} \tag{II.6}$$

The isotopic compositions can now be calculated. We restrict ourselves to discussing hydrogen, nitrogen, carbon dioxide and oxygen, these being representative for 2- and 3-atomic gases.

II.1.2.1 Measurement of $^2H/^1H$ in H_2

The case of hydrogen is complicated by the fact that in the ion source H_2 gas combines with a hydrogen ion to form $^1H_3^+$, which also has mass $\langle 3 \rangle$. The necessary correction is obtained from an extrapolation procedure to the point when the ion source gas pressure equals zero, where $^1H_3^+$ cannot be formed any more.

The constants c_2 and c_3 contain the proportionality factors for the ionisation probabilities, the transmission and collection efficiencies of the molecules with masses $\langle 2 \rangle$ and $\langle 3 \rangle$. Because these factors are unequal unity (<1), the isotope ratios obtained from the current ratios have to be compared to those of a common (internationally adopted; see Chapter 5) standard:

$$\frac{(I_3/I_2)_{\text{sample gas}}}{(I_3/I_2)_{\text{reference gas}}} = \frac{(c_3/c_2)([^2H^1H + {}^1H_3^+]/[^1H_2])_{\text{sample gas}}}{(c_3/c_2)([^2H^1H + {}^1H_3^+]/[^1H_2])_{\text{reference gas}}} \tag{II.7}$$

After correction for the H_3^+ contribution and inserting the isotopic ratios, the $^2\delta$ value becomes:

$$^2\delta = \frac{(^2H/^1H)_{\text{sample}}}{(^2H/^1H)_{\text{reference}}} - 1 = \frac{(I_3/I_2)^0_{\text{sample}}}{(I_3/I_2)^0_{\text{reference}}} - 1 \tag{II.8}$$

where the superscript 0 refers to the correction procedure for the H_3^+ contribution to mass ⟨3⟩ ions.

The absolute $^2H/^1H$ value for the water standard (reference sample) VSMOW is:

$$^2R(\text{VSMOW}) = (^2H/^1H)_{\text{VSMOW}} = 155.76 \times 10^{-6} \tag{II.9}$$

II.1.2.2　*Measurement of $^{15}N/^{14}N$ in N_2*

The mass spectrometric analysis of N_2 for obtaining $^{15}N/^{14}N$ in nitrogen containing materials is equally simple, since in that case we do not have the pressure effect in the ion source that resulted in the formation of $^1H_3^+$ with hydrogen. The measured ion beam ratios are simply translated into isotope ratios:

$$\frac{(I_{29}/I_{28})_{\text{sample}}}{(I_{29}/I_{28})_{\text{reference}}} = \frac{c_{29}/c_{28}([^{15}N^{14}N]/[^{14}N_2])_{\text{sample}}}{c_{29}/c_{28}([^{15}N^{14}N]/[^{14}N_2])_{\text{reference}}}$$

$$= \frac{(^{15}N/^{14}N)_{\text{sample}}}{(^{15}N/^{14}N)_{\text{reference}}} = {}^{15}\delta + 1 \tag{II.10}$$

Air N_2 serves as the reference or standard, and shows very little variation:

$$^{15}R(\text{standard air } N_2) = (^{15}N/^{14}N)_{\text{airN2}} = 0.0036765 \tag{II.11}$$

The mass spectrometric precision obtained is about 0.02‰. However, the laboratory samples handling to prepare the N_2 gas required for the IRMS measurement results in much larger uncertainties, because the chemistry involved is less straightforward.

A set of reference materials is available for checking the overall chemical and analytical procedures (see Table 11.3 in Volume I of the UNESCO/IAEA series) from:

NIST = National Institute of Standards and Technology, Atmospheric Chemistry Group, B364, Building 222, Gaithersburg, MD 20899, United States of America

IAEA = International Atomic Energy Agency, Analytical Quality Control Services, Agency's Laboratory, P.O.Box 100, A-1400 Vienna, Austria, fax +43-1-26007

II.1.2.3 *Measurement of $^{13}C/^{12}C$ and $^{18}O/^{16}O$ in CO_2*

Measurement of the isotopic composition of CO_2 is more complicated because of the abundance of molecules with equal mass 45 and 46, though containing different isotopes.

From Eq. II.6 we can derive the following relations for the isotopic ratios:

$$\frac{I_{45}}{I_{44}} = (c_{45}/c_{44})\frac{CO_2 \text{ with mass } 45}{CO_2 \text{ with mass } 44} = (c_{45}/c_{44})^{45}R^m \tag{II.12a}$$

and

$$\frac{I_{46}}{I_{44}} = (c_{46}/c_{44})\frac{CO_2 \text{ with mass } 46}{CO_2 \text{ with mass } 44} = (c_{46}/c_{44})^{46}R^m \tag{II.12b}$$

II.1.2.3.1 Comparison with reference or standard

As with hydrogen and nitrogen, the ionisation and transmission efficiencies of the isotopic molecules and ions are not 100%, so that the values for the proportionality factors c in Eq. II.6 cannot be replaced by 1. This necessitates comparison of the ion beam ratios with those of a reference or standard.

Laboratories routinely use their own working reference gas (often referred to as the 'working standard' or machine standard, which finally has to be calibrated against the international standard material). Eq. II.11 applies to the sample (s) and the working reference gas (w); we eliminate the efficiency factors c for the mass ⟨45⟩ and mass ⟨46⟩ measurements, and other (generally proportional or multiplicative) instrumental errors:

$$\frac{(I_{45}/I_{44})_s}{(I_{45}/I_{44})_w} = \frac{^{45}R_s^m}{^{45}R_w^m} \tag{II.13a}$$

and

$$\frac{(I_{46}/I_{44})_s}{(I_{46}/I_{44})_w} = \frac{^{46}R_s^m}{^{46}R_w^m} \tag{II.13b}$$

II.1.2.3.2 Isotopic corrections

As is shown in Fig. II.2, the mass ⟨45⟩ and ⟨46⟩ ion beams contain different isotopic ions:

$$\begin{aligned}
I_{44} &= [^{12}C^{16}O^{16}O] \\
I_{45} &= [^{13}C^{16}O^{16}O] + [^{12}C^{17}O^{16}O] \\
I_{46} &= [^{12}C^{18}O^{16}O] + [^{13}C^{17}O^{16}O] + [^{12}C^{17}O^{17}O]
\end{aligned} \tag{II.14}$$

This leads to the isotopic ratios:

$$^{45}R = \frac{I_{45}}{I_{44}} = \frac{[^{13}C^{16}O^{16}O] + [^{12}C^{17}O^{16}O]}{[^{12}C^{16}O^{16}O]} = {}^{13}R + 2\,^{17}R \tag{II.15a}$$

$$^{46}R = \frac{I_{46}}{I_{44}} = \frac{[^{12}C^{18}O^{16}O] + [^{13}C^{17}O^{16}O] + [^{12}C^{17}O^{17}O]}{[^{12}C^{16}O^{16}O]}$$

$$= 2\,^{18}R + 2\,^{13}R\,^{17}R + {}^{17}R^2 \tag{II.15b}$$

The reason for the factors of 2 is that two atoms in the CO_2 molecule can be replaced by the isotope. Translating ^{45}R and ^{46}R into $^{13}C/^{12}C$, $^{17}O/^{16}O$ and $^{18}O/^{16}O$, we are thus dealing with three *unknowns*. For solving the equations we thus need three *givens*, but we can only determine two ratios, namely ^{45}R and ^{46}R.

The third *given* is found in the theoretical relation between the ^{17}O and ^{18}O abundance in natural oxygen (Eq. 2.33):

$$\frac{^{17}R_s}{^{17}R_w} = \sqrt{\frac{^{18}R_s}{^{18}R_w}} \tag{II.16}$$

where s and w refer to a sample and the reference ('laboratory working standard'), respectively. The values for the laboratory reference are known.

The above equations for sample and standard are solved by an iterative procedure:

Suppose $^{17}R = 0$, calculate ^{13}R from ^{45}R, and calculate ^{18}R from ^{46}R; calculate ^{17}R from ^{18}R (Eq. II.16) and now ^{13}R again from ^{45}R using Eq. II.15b; with these values of ^{13}R and ^{17}R calculate again ^{18}R, and repeat.

The values thus obtained for the isotope ratios of the sample $^{13}R_S$ and $^{18}R_S$ are relative to $^{13}R_w$ and $^{18}R_w$ for the working standard. These now have to be adjusted to the VPDB or VSMOW scales for ^{13}C and ^{18}O. The δ values of the sample relative to the international standard are finally calculated as follows:

$$^{\otimes}\delta_{s/r} = \frac{^{\otimes}R_s}{^{\otimes}R_r} - 1 = \frac{^{\otimes}R_s}{^{\otimes}R_w}\frac{^{\otimes}R_w}{^{\otimes}R_r} - 1 = (1 + {}^{\otimes}\delta_{s/w})(1 + {}^{\otimes}\delta_{w/r}) - 1 \tag{II.17}$$

where $^{\otimes}R$ refers to ^{13}R, ^{18}R as well as ^{17}R, and $\delta_{w/r}$ values are obtained by measuring the working standard directly versus the international standards. R_{ref} values are given in Table II.1.

II.1.2.4   *Measurement of $^{18}O/^{16}O$ and $^{17}O/^{16}O$ in O_2*
This analysis includes two measurements, as with CO_2, that is, the ion beam ratios for masses 34/32 and 33/32, while the number of unknowns is also two: $^{18}O/^{16}O$ and $^{17}O/^{16}O$.

Table II.1.   Absolute isotope ratios needed for the calculation procedure of calibrated isotope ratios.

Reference material	$^{13}R \times 10^2$	$^{17}R \times 10^4$	$^{18}R \times 10^3$	$^{2}R \times 10^6$
VSMOW-water		3.790	2.0052	155.76
VSMOW-CO_2		3.867	2.0878142	
VPDB-calcite	1.12372000	3.788666	2.0671607	
VPDB-CO_2	1.12372000	3.808033	2.0883491	
NBS19-calcite	1.12591025	3.784496	2.0626129	
NBS19-CO_2	1.12591025	3.803842	2.0837547	

A similar procedure is previously chosen to calculate the isotope ratios:

$$^{34}R^{\mathrm{m}} = \frac{I_{34}}{I_{32}} = (c_{34}/c_{32})\frac{[^{18}O^{16}O]+[^{17}O_2]}{[^{16}O_2]} = (c_{34}/c_{32})(2^{18}R+^{17}R^2) \quad \text{(II.18a)}$$

and

$$^{33}R^{\mathrm{m}} = \frac{I_{33}}{I_{32}} = (c_{33}/c_{32})\frac{[^{17}O^{16}O]}{[^{16}O_2]} = (c_{33}/c_{32})(2^{17}R) \quad \text{(II.18b)}$$

After correction for instrumental errors (specific and probably time-dependent for each instrument), the corresponding δ values are:

$$^{34}\delta = \frac{^{34}R^{\mathrm{m}}_{\mathrm{sample}}}{^{34}R^{\mathrm{m}}_{\mathrm{reference}}} - 1 = \frac{^{34}R_{\mathrm{s}}}{^{34}R_{\mathrm{r}}} - 1 = \frac{2^{18}R_{\mathrm{s}}+^{17}R_{\mathrm{s}}^2}{2^{18}R+^{17}R_{\mathrm{r}}^2} - 1 \quad \text{(II.19a)}$$

and

$$^{33}\delta = \frac{^{33}R^{\mathrm{m}}_{\mathrm{sample}}}{^{33}R^{\mathrm{m}}_{\mathrm{reference}}} - 1 = \frac{^{33}R_{\mathrm{s}}}{^{33}R_{\mathrm{r}}} - 1 = \frac{2^{17}R_{\mathrm{s}}}{2^{17}R_{\mathrm{r}}} - 1 = \frac{^{17}R_{\mathrm{s}}}{^{17}R_{\mathrm{r}}} - 1 \quad \text{(II.19b)}$$

Because the $^{17}R_{\mathrm{r}}$ and $^{18}R_{\mathrm{r}}$ values for the reference samples or standards are known, $^{17}R_{\mathrm{s}}$ and $^{18}R_{\mathrm{s}}$ and thus the $^{17}\delta$ and $^{18}\delta$ values, can be easily calculated.

II.2   RADIOMETRY FOR RADIOACTIVE ISOTOPES

A number of instruments are available for measuring radioactivity, each with their specific advantages and disadvantages. The main instruments discussed here are (i) gas counters and (ii) liquid scintillation spectrometers.

   With both techniques the radioactive sample is measured as an *internal source*, that is, it is mixed with, or is an intrinsic part of the counting medium responsible for the operation of the instrument. This has the following advantages:

1. The β particles from the radioactive decay do not need to penetrate the wall of the instrument; this is especially relevant for detecting β^- particles from ^3H and ^{14}C decay, as in both cases the β energy is very small.
2. The counting efficiency is nearly 100%, and the detection angle is 4π.

II.2.1   Gas counters

Although the operation of the various gas counter types is different, their basic construction is the same; the principle is shown in Fig. II.3.

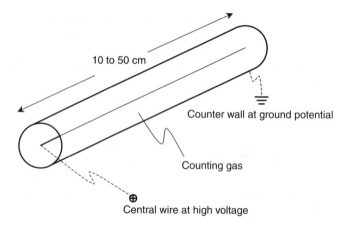

Figure II.3.   Cross-section of a gas-filled counter tube for detection of radioactive radiation. The ionisation phenomena in the counter are shown in the cross-section of Fig. II.4.

Basically there are three different types of gas counters, dependent on the magnitude of the high-voltage applied between the wire and the counter wall:

1.   ionisation chamber
2.   proportional counter
3.   Geiger-Müller (GM) counter.

The counters are filled with a counter gas that has suitable ionisation characteristics.

II.2.1.1   *Ionisation chamber*

The incoming radiation, that is a particle originating from a decay reaction inside or outside the counter tube, loses its energy by collisions with gas molecules in the counter. These events cause the molecules to become ionised. In the electric field inside the tube, the heavy positive ions move slowly towards the wall and become neutral by picking up electrons. The small, negative *primary electrons* set free by the collision become accelerated towards the positive central wire. After being collected by the wire, these electrons are observed as a small electric current or an electric pulse in the wire. In the electronic circuitry attached to the counter, this pulse is registered and the pulse size measured. This is representative for the number of primary electrons, and thus of the energy of the incoming particle. In this way information is obtained about the radioactive decay event.

II.2.1.2   *Proportional counter*

The primary electrons are accelerated in the electric field between the wire and counter wall. If the voltage on the counter wire is sufficiently high the primary electrons, accelerated towards the wire, obtain enough velocity and energy to ionise other gas molecules and so create more free electrons. The same is true for these new electrons, and so on. In this way an avalanche of electrons is created that moves towards and is collected by the wire (Fig. II.4). This phenomenon is called *gas multiplication*. The resulting electric pulse is proportional to this gas multiplication (and thus depends on the high voltage), and is still *proportional to the number of primary electrons and thus to the original energy of the incoming particle* (PGC).

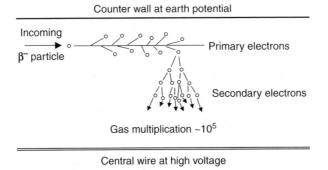

Figure II.4. Lateral cross-section through a gas counter. An incoming β⁻ particle loses its energy through primary ionisation. At low voltage only these electrons reach the wire, resulting in a small electric pulse (ionisation chamber). If the voltage is sufficiently high, the number of primary electrons can be multiplied by a factor of 10^5 (PGC). At very high voltage the secondary ionisation is maximal, and the electron avalanche occupies the entire length of the counter: the pulse height no longer depends on the incoming particle energy or the voltage (GM counter).

II.2.1.3 *Geiger-Müller counter*

During operation as a proportional counter the secondary electron avalanche is very local. At very high voltage, however, the secondary ionisation process extends over the entire length of the counter tube. The pulse size is maximal and no longer depends on the energy of the incoming particle or the wire voltage: in the GM counter the existence of the particle is registered, but cannot be recognised, as in the case of the proportional counter. This is essential, however, in order to be able to distinguish the particle (e.g. a β⁻ particle from ^{3}H or ^{14}C) from strange particles.

Our conclusion, therefore, has to be that the only suitable gas counter for our purpose is the one which operates as a proportional counter. This equipment was the first established for measuring the radioactivities of ^{3}H and ^{14}C; these atoms have to be brought into a gaseous form suitable for operation as counting gas in a counter. This is essential, because the β radiation cannot penetrate the counter wall as the β energies are extremely low (Fig. 6.2). Later, another technique came into use; the liquid scintillation spectrometer.

II.2.1.4 *Counter operation*

At both ends of the PGC the geometry is distorted, as is the electric field. The detection of particles from radioactive decay is, therefore, less efficient. The consequence is that the overall *counter efficiency* is not exactly known, and routinely an absolute measurement is not feasible. The solution to this problem is to perform a relative measurement: the sample is measured under exactly the same conditions as a reference sample or, ultimately, as an international standard (Section 6.1.2). Alternatively, the dependence of the detected radioactivity of a sample on the various parameters has to be exactly known. Without going into detail, we mention the essential parameters.

The amount of gas in the counter: since sample and reference are being measured in the same counter the counter volume (V) is not relevant, so that measuring the pressure (p) and temperature (T) is sufficient: the amount of gas = Vp/T with $T = t$ (°C) + 273.15 K.

The purity of the counting gas: if the gas contains electro-negative impurities, that is, gas molecules with a strong attraction to electrons, part of the primary and secondary electrons are lost, thus decreasing the pulse size and the counting rate.

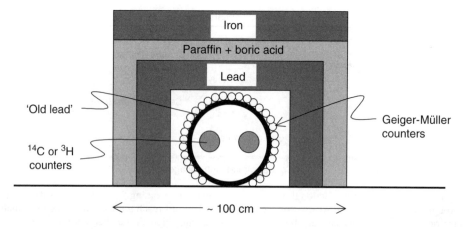

Figure II.5.   Cross-section of a typical counter with shielding against radiation from the surroundings, including cosmic radiation. The GM counters operate in 'anti-coincidence' with the proportional counters in the middle. The low-mass hydrogen atoms in paraffin are very efficient in reducing the (penetrating) energy (speed) of high-energy neutrons. Boron has a high absorbing capacity for low-energy neutrons. The high-mass lead absorbs γ radiation. 'Old lead' has lost the radioactive lead isotope ^{210}Pb (RaD), due to its short half-life (22 years).

The background of the counter: that is, the counting rate (number of counts per second) of the counter without a radioactive source, in our case CO_2 and C_2H_2, or H_2 without ^{14}C or ^{3}H, respectively. This background has to be subtracted from the overall counting rate. In order to minimise the background – which would otherwise be orders of magnitude larger than the actual (net) counting rate we are interested in – the counter is made of low-activity materials and shielded against radiation from the surroundings (see Fig. II.5).

Most counters are made of copper or quartz with volumes ranging between 0.5 and a few L, although mini-counters also exist. An international summary was published by Mook (1983).

In principle, the data handling from radiometric equipment is simple. After a certain measuring time t_m, a number of N_t disintegrations in the counter have been observed. The statistical uncertainty or *standard deviation* of this number is $\sqrt{N}$.

The gross counting rate of N_t/t_m (counts per second) must be diminished by the background B (also in cps) to give the net counting rate A':

$$A'(t) = N_t/t_m - B \tag{II.20}$$

This result has then to be standardised, that is, corrected to a standard pressure and temperature for the sample gas in the counter.

We have then obtained the standardised absolute activity: ^{3}A for ^{3}H, ^{14}A for ^{14}C (Section 5.1.2).

The internationally established standard (Oxalic Acid) must be measured under exactly the same conditions. The activity ratio can then be calculated as that between these two activities (Eq. 6.4):

$$^{14}a = {}^{14}A_\text{sample} / {}^{14}A_\text{standard} \tag{II.21}$$

The standard deviation in the measured activity is (standard in B is negligibly small):

$$\sigma_\text{A} \approx \sigma_\text{N} = \frac{1}{t_\text{m}}\sqrt{N} = \sqrt{\frac{A+B}{t_\text{m}}} \tag{II.22}$$

II.2.2   Liquid scintillation spectrometer

II.2.2.1   *Physical principle*

This method is based on the principle that certain materials emit light after their molecules have been excited by collisions with high-energy particles.

This process, called *luminescence*, takes place in a solid or liquid. Our interest focuses on the use of liquids, because there are methods to transfer carbon (including ^{14}C) to benzene which – with some precautions – is suitable as a *scintillator*. Water (with ^{3}H) can be mixed with a scintillating liquid, a commercially available *cocktail* of various constituents. The requirement for this mixture is that it mixes well with water, transfers the primary electron energies into fluorescent energy, and finally into low-energy fluorescent light within the secondary-emission range of the photo-electrode of the photomultiplier.

In this way we are again dealing with internal radioactive sources, as with the gas counters. For higher-energy β and γ particles, solid crystals are available made of organic (anthracene) or inorganic compounds (sodium iodide).

The result is that each incoming particle causes a light flash in the scintillator which can be 'seen' by a light sensitive detector, the *photocathode*. The fact is that certain materials emit electrons when they are hit by a light flash, the so-called *photoelectric effect*; the number of *primary electrons* is very small. Therefore, as with the proportional counter operation, an 'multiplication' process is needed. This is realised in a *photomultiplier tube* (Fig. II.6).

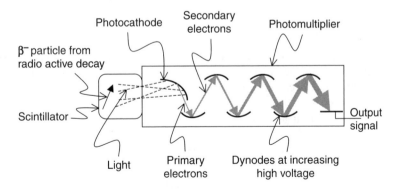

Figure II.6.   Schematic drawing of a (liquid) scintillation counter, the detector part of the LSS. An electron from β⁻ decay in the scintillating crystal or liquid causes a light flash that is 'transformed' into electrons at the photocathode; the number of primary electrons is multiplied by secondary emitting dynodes at increasing high voltage.

Each primary electron is accelerated and able to cause the emission of additional *secondary electrons* in the second electrode plate or *dynode* at a higher positive voltage. This process is repeated several times, so that finally an electric pulse arrives at the last dynode that is easily measurable. Also here is the final pulse proportional to the multiplication factor – a direct function of the voltage – and the energy of the incoming particle. This allows the selective registration of only those particles that originate from the radioactive decay event of interest (such as ^{14}C or ^{3}H).

II.2.2.2   *Counter operation*

In Section II.2.4.2 we have mentioned the preparation of benzene from CO_2 via acetylene. Typical quantities being treated are 1–10 grams of carbon, resulting in 1.2–12 mL of benzene. Since the scintillation light emitted by excited C_6H_6 does not have the frequency for which the photocathode in the multiplier tube is sensitive, certain high-molecular compounds are added in order to shift the frequencies to lower energies.

Generally, a large number of samples is arranged in an LSS or counter, in which the samples are in turn placed in front of the photomultiplier automatically. Also here arrangements have been made to reduce the background radiation as much as possible, primarily by using an anti-coincidence arrangement of other scintillation counters (instead of the use of GM counters with PGC).

The advantage of the LSS is that it can be commercially purchased as an automated instrument, while a gas counter set-up has to be made by the laboratory. On the other hand, from a physical point of view the operation of a PGC is more 'transparent'.

In principle, the same type of calculations lead to the activity of the sample, as shown for the gas counters (Eqs II.21 and II.23). Related to the problem of gas purity with the gas counters, liquid scintillation may suffer from impurities in the benzene plus scintillator mixture. This results in a decrease in the size of the electrical output pulse. A correction needs to be applied for this effect of *quenching*.

II.3   MASS SPECTROMETRY FOR LOW-ABUNDANCE ISOTOPES

II.3.1   Physical principle of AMS

Detecting the presence of isotopes with a very low abundance such as the radioactive isotopes of ^{14}C in carbonaceous material, wood, charcoal, peat, bone, shells or groundwater, by detecting the radioactive decay is very inefficient. Of all ^{14}C present in a sample, only one thousandth decays in about eight years or 3.8×10^{-12} per sec.

A much more efficient detection method is mass spectrometry; that is, measuring the actual concentration of ^{14}C. Nevertheless, only in the late 1970s have attempts been successful to detect single ^{14}C particles in carbon and separate these from other foreign elements.

IRMS operates at a few kV and accelerates molecules such as CO_2, which are separated into ion beams of masses 44, 45 and 46. For $^{14}CO_2$ the mass is 46, the signal for which – due to its extremely low abundance of $<10^{-12}$ – is completely lost in the 46 signal from the various stable isotopic combinations in CO_2. It appears that by making the step to C ions instead of CO_2 ions, the 'isobar' problem can be solved. This means accelerating over MV rather than kV.

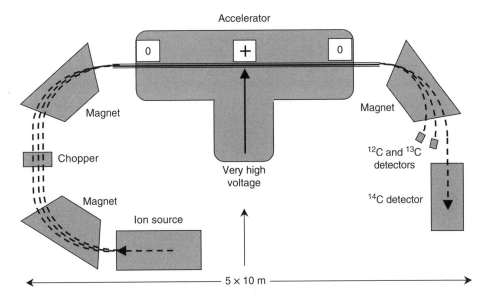

Figure II.7. Schematic and simplified drawing of an AMS (instrument manufactured by High Voltage Engineering, Amersfoort, the Netherlands). The ion source produces negative carbon ions. In order to reduce the overwhelming contribution of ^{12}C, the number of C ions are artificially reduced by a factor of 100 in the chopper. The C^- ions are accelerated in the accelerator tube in the left half of the accelerator by an electric field of about 3 MV. In the middle they are stripped of electrons to C^{3+}, so that in the right half they are further accelerated as positive ions. In the final magnet the ^{12}C, ^{13}C and ^{14}C ion are separated and finally counted.

Registration becomes possible in a nuclear particle detector under these high-voltage conditions. Contrary to IRMS, single particles can then be detected rather than beam currents.

For this purpose existing nuclear accelerators have been used, specifically Tandem Van de Graaff Accelerators. Specific, dedicated machines were later constructed and put into operation that are much smaller (Fig. II.7).

The complete set-up of the accelerator with supporting equipment is called an *Accelerator Mass Spectrometer* (AMS). The essential features of an AMS are as follows:

1.   The ion source produces negative ions. This is necessary because contamination of carbon by nitrogen cannot be avoided and a N^+ ion beam would make the detection of ^{14}C practically impossible, ^{14}C having the same position in the ultimate mass spectrum. Negative ions, on the other hand, have the advantage that N^- ions are extremely unstable and lose their charge before entering the accelerator.
2.   The negative ions are produced by bombarding the sample graphite with Cs atoms. At present only very few CO_2 ion sources are operational for AMS. This would be less demanding for the chemical preparation procedures, but such a source is a complex device.
3.   Ultimately, positive ions are easier to detect, so that inside the accelerator the negative electric charge of the carbon ions is removed and replaced by a positive charge by steering the ions through a *stripper gas* (Ar). The stripping process especially

destroys molecular isobars with mass 14, such as the beam contaminants ^{13}CH and ^{12}CH. The now positive ions are again accelerated and subsequently mass selected in a magnetic field so that ^{12}C, ^{13}C and ^{14}C can be detected separately.
4. Before entering the accelerator, the ions are mass separated by a magnetic field and the ^{12}C beam is reduced by a factor of 100 by a fast rotating diaphragm (chopper, Fig. II.7). The second magnet recombines the three ion beams.

II.3.2 Advantages and disadvantages

The particular *advantages* of the AMS technique, in comparison with radiometry, are:

1. Only milligram (or smaller) quantities of carbon are needed for one analysis, compared to grams for the radiometric techniques (see Table II.2).
2. The analysis is considerably faster; it lasts an hour instead of days.

The *disadvantages* are that:

1. The capital investment is almost 100 times greater.
2. The AMS technique is more complicated from a technical point of view, and needs more manpower to keep it in operation.
3. The precision is not yet as high as the best counters (PGC or LSS) (see Table II.2).

II.4 DETERMINATION OF ^3H THROUGH MASS SPECTROMETRIC MEASUREMENT OF ^3He

Very low ^3H abundances (theoretically down to 0.0005 TU) can be measured by thoroughly degassing and storing 50–100 mL of water sample for at least half a year in a tightly sealed container, and subsequently recollecting the decay product of tritium, ^3He. The amount of this inert gas can be quantitatively measured by mass spectrometer. The ^3He produced during this storage time (t) is (Section 4.3.4 and Eq. 9.7):

$$^3He_t = {}^3H_0 - {}^3H_t = {}^3H_0(1 - e^{-\lambda t})$$

so that the original ^3H content of the sample was:

$$^3H_0 = {}^3He/(1 - e^{-\lambda t}) \tag{II.23}$$

To give an impression: the amount of ^3He formed during half a year storage of 100 mL of water with a ^3H concentration of 0.5 TU is:

$$^3He_t = 0.5 \times 10^{-18} \times 2 \times (100/18)(1 - e^{-(\ln 2)/12.43)\times 0.5})22.41 \times 10^3$$
$$= 3.4 \times 10^{-15} \text{ mL}$$

It is obvious that for detecting such a small amount of gas, a mass spectrometer is needed.

Table II.2.   Summary of specific features for the various isotopic techniques. The ^{3}H and ^{14}C precisions quoted are valid for samples of approximately 10 TU and present-day activities, respectively. (The asterisks refer to cases where larger quantities of sample allow greater precision. For the radiometric techniques we have assumed a counting time of one or two days.)

	Rare isotopes	Medium type	Medium quantity	Precision
IRMS	^{2}H, ^{13}C, ^{18}O (^{15}N, ^{34}S)	H_2, CO_2, N_2	few mL gas to few 0.1 mL gas	^{2}H: 0.1‰ ^{13}C: 0.01‰ ^{18}O: 0.02‰ ^{15}N: 0.02‰
PGC	^{3}H	H_2, C_2H_6 with enrichment	few mL H_2O few 100 mL H_2O	1 TU (NE) 0.1 TU (E)
	^{14}C	CO_2, C_2H_2, C_3H_8	1–50 L CO_2 (equiv.)	0.5–0.1%*
LSS	^{3}H	H_2O	few mL H_2O few 100 mL H_2O	few TU (NE) few 0.1 TU (E)
	^{14}C	C_6H_6	few to 15 mL C_6H_6	0.5–0.2%*
AMS	^{14}C	C (graphite) (CO_2)	0.1–2 mg C few mL CO_2	0.5%* ?

*For carbon with $^{14}a = 100\%$. IRMS = Isotope Ratio Mass Spectrometer, PGC = Proportional Gas Counter, E = Artificially enriched, generally by electrolysis, LSS = Liquid Scintillation Spectrometer, AMS = Accelerator Mass Spectrometer, NE = not enriched.

II.5   REPORTING ^{14}C ACTIVITIES AND CONCENTRATIONS

II.5.1   The choice of variables

The following definitions are in use:

The *absolute (specific)* ^{14}*C activity*, that is the ^{14}C radioactivity (in Bq or, conventionally, in dpm per gram of carbon) is given the symbol

$$^{14}A \text{ (in dpm/gC)} \tag{II.24}$$

In general, ^{14}C laboratories are not able to make an absolute measurement because the measuring efficiency is unknown. Also, in general, the absolute ^{14}C content of a sample is not relevant. The sample activities are therefore compared with the activity of a reference material, the international standard. In reality, the number of ^{14}C registrations (β registrations from ^{14}C decay in radiometric detectors such as proportional counters and liquid scintillation counters, registrations of ^{14}C concentration in AMS systems) are related to the number of registrations from the reference sample under equivalent conditions. This results in the introduction of a ^{14}*C activity ratio* or ^{14}*C concentration ratio*:

$$^{14}a = \frac{\text{measuring efficiency} \times ^{14}A_{\text{sample}}}{\text{measuring efficiency} \times ^{14}A_{\text{reference}}} = \frac{^{14}A_{\text{sample}}}{^{14}A_{\text{reference}}}$$

$$= \frac{^{14}\text{C decay rate in the sample}}{^{14}\text{C decay rate in the ref. material}} = \frac{^{14}\text{C concentration in the sample}}{^{14}\text{C concentration in the ref. material}} \tag{II.25}$$

Because the detection efficiencies are equal for sample and standard, in the numerator and denominator of the last two fractions,

the use of the ratio ^{14}a is adequate for any type of measuring technique. Henceforth, the symbol ^{14}A will be used for the ^{14}C *content*, radioactivity as well as concentration,

whether the analytical technique applied is radiometric or mass spectrometric.

Under natural circumstances the values of ^{14}a are between 0 and 1. In order to avoid a large number of decimals, it is general practice to report these values in % (per cent), which is equivalent to $\times 10^{-2}$. Therefore, the factor 10^2 should *not* enter into equations (as $^{14}a/10^2$).

In some cases the differences in ^{14}C content between samples are small. Therefore, the use of relative abundance has been adopted from the stable-isotope field, *in casu* the *relative ^{14}C content (activity or concentration)*, $^{14}\delta$, defined as the difference between sample and standard ^{14}C content as a fraction of the standard value:

$$^{14}\delta = \frac{^{14}A - {}^{14}A_{\text{ref}}}{^{14}A_{\text{ref}}} = \frac{^{14}A}{^{14}A_{\text{ref}}} - 1 = {}^{14}a - 1 \tag{II.26}$$

The values for δ are small numbers and are therefore generally given in ‰ (per million), which is equivalent to $\times 10^{-3}$. Thus, the factor 10^3 should *not* enter mathematical equations (as $\delta/10^3$).

A ^{14}C *reference material* or *standard* was chosen to represent, as closely as possible, the ^{14}C content of carbon in naturally growing plants. The ^{14}C content of the standard material itself does not need to be, in fact is not, equal to the *standard ^{14}C content*. The definition of the standard ^{14}C activity is based on the specific activity of the original NBS oxalic acid (Ox1), as will be discussed in more detail.

For a more detailed treatment of the standardisation and definitions for a number of ^{14}C variables, the reader is referred to section 11.5 of Volume I from the UNESCO/IAEA series. We restrict ourselves to a brief summary, especially applicable in hydrological research.

Volume I, chapter 11 of the UNESCO/IAEA series gives a number of examples relating to calculation procedures for the various fields which apply ^{14}C activities. Here we restrict ourselves to the aquatic applications, namely hydrology and oceanography.

II.5.2   Special case 1: hydrology

From a geochemical point of view, it is the most meaningful to give the ^{14}C content of a water sample as ^{14}a – not corrected for isotope fractionation and valid for *the time of sampling* – rather than a δ value. Instead of applying a normalisation correction, the initial ^{14}C content of groundwater is determined by specific geochemical reasoning of the origin of the inorganic carbon content. Furthermore, when dealing with groundwater ages it is irrelevant – from a hydrological point of view – whether ages count back in time from the year of sampling (calculation based on $^{14}a^S$) or from 1950 (calculation based on $^{14}a = {}^{14}a^0$). Moreover, the precision of routine ^{14}C dating is in any case ± 50 years or more. Consequently, we can equally well use the most straightforward ^{14}a value in % as is obtained in simple laboratory procedures.

> These values are often given in *per cent of modern carbon* (pMC) or *percent modern* (pM). In addition, pM/100 ($= {}^{14} a^S$) is sometimes called *fraction modern*. However, the symbol pM is in use by water chemists and oceanographers as picoMole. Therefore, pmc, pMC, pM and similar variants should not be used: % is adequate in combination with a well-defined symbol. Nevertheless, the use of pMC has become so established that this 'unit' is difficult to avoid. In this respect it seems more reasonable to use the symbol %MC instead of pMC.

Example: groundwater

The ^{14}C result communicated by the laboratory is generally the normalised value (i.e. corrected for isotope fractionation). For hydrological applications, however, we have to use the un-normalised value; the ^{14}C result obtained from the laboratory therefore needs to be denormalised (the reverse of Eq. 6.7).

We take an example with a measured and normalised activity or concentration ratio:

$$^{14}a_N = 0.537 \; (= 53.7\%)$$

and a measured $^{13}\delta$ value of the total carbon content (as is obtained by the extraction procedure)

$$^{13}\delta = -13.82\%\!o$$

In groundwater hydrology we are interested in the ^{14}C content of the water sample at the time of sample collection. The ^{14}a value therefore has to be de-normalised (the equation numbers in square brackets refer to section 11.5 of Volume I in the UNESCO/IAEA series):

$$^{14}a = {}^{14}a_N[(1 + {}^{13}\delta)/0.975]^2 = 0.549 = 54.9\% \qquad \text{[cf. 11.40]}$$

The ^{14}C content in the year of sampling (e.g. 1998) is then:

$$^{14}a^S = {}^{14}a_N[(1 + {}^{13}\delta)/0.975]^2 \exp[-(1998 - 1950)/8267] = 0.546 = 54.6\%$$
$$(\equiv 54.6 \text{ pMC} = \% \text{ of modern carbon} = \%MC) \qquad \text{[cf. 11.43]}$$

Using more or less sophisticated models, the ^{13}C and ^{14}C data, together with information on the chemical composition of the sample, can be used to estimate the sample age (i.e. the period of time since the infiltration of the water). A straightforward 'water age' as would be obtained by simply applying Eq. II.28 is not possible.

II.5.3   Special case 2: oceanography

The same equation holds for the oceanographic applications. However, as the spread of the data generally is quite small, it is common practice to report the ^{14}C data as relative numbers, $^{14}\delta$ values (cf. Eq. II.26):

$$^{14}\delta^S = {}^{14}a^S - 1 = \frac{^{14}A}{^{14}A_r} - 1$$

In oceanographic studies, the ^{14}C data have to be corrected for isotope fractionation (= Normalised to $^{13}\delta = -25\%o$) according to Eq. 6.7):

$$^{14}a_N = {}^{14}a\left[\frac{1 + {}^{13}\delta_N}{1 + {}^{13}\delta}\right]^2$$

The ^{14}C data also have to be corrected for radioactive decay between 1950 and the year of sampling (S refers to the time of sampling $\approx$ time of measurement). The fact is that routine measurement results in a ^{14}C content valid for the year of definition of the reference material (Ox), that is, 1950, because the sample and the reference material both decay at the same rate:

$$^{14}a^S = {}^{14}a\,e^{-\lambda(t_S - 1950)} = {}^{14}a e^{-(t_S - 1950)/8267}$$

(cf. Eq. 4.20; the sample activity has diminished since 1950).
The overall equation is now:

$$^{14}\delta_N^S = {}^{14}a_N^S - 1 = {}^{14}a \cdot e^{-(t_S - 1950)/8267}\left(\frac{0.975}{1 + {}^{13}\delta}\right)^2 - 1 \qquad (\text{II.27})$$

These equations are used to express the ^{14}C content of samples of oceanwater.

Example: oceanic DIC
The ^{14}C content resulting from a routine measurement includes normalisation to $^{13}\delta = -25\%o$: deep-ocean bottom water DIC:

$$^{14}a = 0.9201 = 92.0\% \quad \text{and} \quad {}^{13}\delta = +1.55\%$$

$$^{14}a_N = {}^{14}a\left[\frac{1 + {}^{13}\delta_N}{1 + {}^{13}\delta}\right]^2 = 0.9201\left[\frac{0.975}{1.00155}\right]^2 = 0.872 = 87.2\%$$

from which: $^{14}\delta_N = {}^{14}a_N - 1 = -0.128 = -128\%o$ [cf. 11.42]

When corrected for the fact that the resulting ^{14}C content is valid for the year 1950 instead of the year of sampling (e.g. 1990):

$$^{14}\delta_N^S = {}^{14}a_N \exp[-(1990 - 1950)/8267] - 1 = -0.132 = -132\%o$$

In literature ^{14}C data thus obtained are generally denoted by Δ.

II.5.4 ^{14}C ages

In geological and archaeological dating, the ages used are the *conventional ^{14}C ages*. By international convention, the definition of the conventional ^{14}C age is based on (Section 6.1.4):

1. The *initial ^{14}C activity* (i.e. the activity of the sample material during 'growth' is equivalent to the standard activity in AD 1950.
2. The ^{14}C activities have to be normalised for fractionation (samples to $^{13}\delta = -25‰$, Ox1 to $-19‰$, Ox2 to $-25‰$) (see Appendix I).
3. The original (Libby) half-life of 5568 has to be used.

The ages are then calculated by applying:

$$\text{Conventional age} = -8033\ln{}^{14}a_N \qquad\qquad\text{(II.28)}$$

This defines the ^{14}C time scale in years bp (Before Present, i.e. before AD 1950).

This time scale needs to be calibrated in order to obtain historical (*calibrated*) ages (calAD, calBC, calBP). For the calibration procedures and conventions we refer to the special Calibration Issues, published by the *Radiocarbon Journal*.

II.5.5   Summary

The 'rational' symbols defined and discussed in Volume I of the series are based on:

- ^{14}A = absolute radioactivity (in Bq/gC or in dpm/gC).
- ^{14}a = activity or concentration ratio between sample and standard (in %).
- $^{14}\delta$ = relative difference in activity or concentration (in ‰).

Corrections are introduced for the effects of isotope fractionation and radioactive decay during the intervals of time between formation of the sample carbon (e.g. plant growth), the sample collection and measurement.

In hydrology we prefer to use the quantity ^{14}a in %.

APPENDIX III

Chemistry of carbonic acid in water

III.1   INTRODUCTION

When studying the carbon isotopic composition of water, whether it concerns fresh water or sea water, a complication arises from the fact that the dissolved inorganic carbon always consists of more than one compound. The presence of gaseous CO_2 and solid calcium carbonate may also be relevant. In fact we are dealing with the following compounds and concentrations:

gaseous CO_2 (occasionally denoted by CO_2g) with a partial pressure P_{CO_2}
dissolved CO_2 (denoted by CO_2aq)
dissolved carbonic acid, H_2CO_3, with $a = [H_2CO_3] + [CO_2aq]$
dissolved bicarbonate, HCO_3^-, with $b = [HCO_3^-]$
dissolved carbonate, CO_3^{2-}, with $c = [CO_3^{2-}]$
total dissolved inorganic carbon, DIC, with $C_T = a + b + c$
solid carbonate, $CaCO_3$ (occasionally denoted by s)

The ambiguity in using carbon isotopes now reduces to two observations:

1. On the one hand, only the isotopic composition of a single compound in relation to that of another is geochemically or hydrologically meaningful. We have seen an example of this statement when discussing the carbon isotopic composition of groundwater, for ^{13}C as well as C^{14} (Figs 5.6, 6.4 and 9.5). In other words, an isotopic fractionation factor is a fundamental, physico/chemical quantity only if it is the ratio between two isotopic ratios of single compounds. For example:

$$^{13}\alpha_{a/b} = {^{13}R_a}/{^{13}R_b} \quad \text{and} \quad ^{13}\alpha_{c/b} = {^{13}R_c}/{^{13}R_b}$$

2. (a) On the other hand, the carbon isotopic composition of a mixture of compounds is relevant in mass balance considerations. For example, if CO_2 or $CaCO_3$ are being removed from a dissolved carbon solution, or for estuarine mixing of fresh- and sea water, the total ^{13}C mass balance has to be taken into account. Some examples are given in this Appendix.
    (b) Measuring the carbon isotopic composition of a solution involves extracting the total CO_2 from the sample as a whole after acidification, instead of single compounds.

For the necessary translation of $^{13}\delta$ for total dissolved carbon to $^{13}\delta$ for single compounds and *vice versa*, the chemistry of the dissolved inorganic carbon is required.

Once the various concentrations of the dissolved species – to be derived in the next sections – are known, the ^{13}C mass balance is:

$$([CO_2aq] + [H_2CO_3] + [HCO_3^-] + [CO_3^{2-}])^{13}R_{DIC} = [CO_2aq]^{13}R_{CO_2aq}$$
$$+ [H_2CO_3]^{13}R_{H_2CO_3} + [HCO_3^-]^{13}R_{HCO_3} + [CO_3^{2-}]^{13}R_{CO_3} \tag{III.1}$$

or, inserting the previously mentioned symbols for the various concentrations and combining the concentration of CO_2aq and the carbonic acid, H_2CO_3, the latter being a negligibly small fraction, the isotopic composition of the total dissolved carbon ($C_T = a + b + c$) is:

$$^{13}R_{C_T} = \frac{aR_a + bR_b + cR_c}{C_T} \tag{III.2}$$

and, conversely, inserting the proper fractionation factors as previously mentioned:

$$^{13}R_b = \frac{C_T \, ^{13}R_{C_T}}{a\alpha_{a/b} + b + c\alpha_{c/b}} \tag{III.3}$$

We devote the following sections to analysing the chemical composition of carbonate waters.

III.2   CARBONIC ACID EQUILIBRIA

Dissolved CO_2 exchanges with CO_2 gas in the presence of gaseous CO_2:

$$CO_2(g) + H_2O \Leftrightarrow CO_2(aq) + H_2O \tag{III.4}$$
$$CO_2(aq) + H_2O \Leftrightarrow H_2CO_3 \tag{III.5}$$

where g and aq refer to the gaseous and dissolved phases, respectively. Although the concentration of $CO_2(aq)$ far exceeds that of dissolved H_2CO_3 (in the order of 10^3), we denote the concentration of all dissolved CO_2 by $[H_2CO_3]$. The equilibrium condition between the phases is quantified by the *molar solubility* K_0 (Henry's law):

$$K_0 = \frac{[H_2CO_3]}{P_{CO_2}} \tag{III.6}$$

where the atmospheric CO_2 partial pressure, P_{CO_2}, is in atm, K_0 is the solubility in mol L^{-1} atm^{-1}, and $[H_2CO_3]$ is the dissolved CO_2 concentration in mol/kg of water.
    H_2CO_3 dissociates in water according to:

$$H_2CO_3^+ \Leftrightarrow H^+ + HCO_3^- \tag{III.7}$$

and

$$HCO_3^- \Leftrightarrow H^+ + CO_3^{2-} \tag{III.8}$$

where the equilibrium conditions are quantified by the *dissociation* or *acidity constants*:

$$K_1 = \frac{[H^+][HCO_3^-]}{[H_2CO_3]} \tag{III.9}$$

and

$$K_2 = \frac{[H^+][CO_3^{2-}]}{[HCO_3^-]} \tag{III.10}$$

Finally, the dissociation of water obeys the equilibrium condition:

$$K_w = [H^+][OH^-] \tag{III.11}$$

Here we must emphasise that although the hydrogen ion, H^+, is commonly hydrated to form H_3O^+, or even to multiple hydrated ions, we write the hydrated hydrogen ion as H^+ since the hydrated structure does not enter the chemical models.

The $[H^+]$ concentration is generally given as a pH value, defined as the negative logarithm:

$$pH = -^{10}log[H^+] \tag{III.12}$$

The total concentration of dissolved inorganic carbon (= total carbon, also denoted by ΣCO_2 or ΣC or DIC) is defined by:

$$C_T = [CO_2aq] + [H_2CO_3] + [HCO_3^-] + [CO_3^{2-}] = a + b + c \tag{III.13}$$

The *alkalinity* is a practical quantity, following from the conservation of electroneutrality in solutions where the metal-ion concentrations (Na, Ca, Mg) and pH are constant:

$$A_T = [HCO_3^-] + 2[CO_3^{2-}] + [OH^-] - [H^+] + [\text{other weak acid anions}] \tag{III.14}$$

in which concentrations of other weak acids may be included in the interest of high precision, such as humic acids in fresh water or borate, $[B(OH)_4^-]$, in sea water.

Under natural conditions, $[H^+]$ and $[OH^-]$ are negligible compared to the carbonate species concentrations. The sum of the weak acid and alkali-ion concentrations, determined by acid titration and referred to as the total alkalinity, thus approximate the *carbonate alkalinity*, defined as:

$$A_C = [HCO_3^-] + 2[CO_3^{2-}] = b + 2c \tag{III.15}$$

If the water contains Ca^{2+} (or Mg^{2+}) and carbonate or is in contact with calcite, the dissociation equilibrium of calcite also affects the carbon chemistry:

$$CaCO_3 \Leftrightarrow Ca^{2+} + CO_3^{2-} \tag{III.16}$$

where the concentrations are limited by the *solubility product*:

$$K_{CaCO_3} = [Ca^{2+}][CO_3^{2-}] \tag{III.17}$$

III.3   THE EQUILIBRIUM CONSTANTS

All values of the solubilities and dissociation constants are basically temperature dependent. However, the K values also depend on the solute concentrations, as the formation of ion complexes between the carbonic ions and molecules and ions in the solution hinder the dissolved carbonic molecules and ions from taking part in the thermodynamic equilibrium reactions. Therefore, in the thermodynamic equation the concentrations have to be replaced by their *activities*, which are smaller than the concentrations. The *thermodynamic solubility constant* is:

$$K_0 = \frac{a_{H_2CO_3}}{P_{CO_2}} = \frac{\gamma_a[H_2CO_3]}{P_{CO_2}} \tag{III.18a}$$

where, in general, the *activity coefficients* $\gamma < 1$ ($\gamma = 1$ for an *ideal solution*, that is, with zero solute concentrations or zero *ionic strength*).

In the non-ideal solutions of sea water and brackish water it is more practical to describe the relation between the real, *measurable* concentrations by the *apparent solubility constant*:

$$K_0' = \frac{[H_2CO_3]}{P_{CO_2}} = \frac{K_0}{\gamma_a} \tag{III.18b}$$

The *thermodynamic* and the *apparent acidity (dissociation) constants* of the first and second dissociation of carbonic acid (Eqs III.6, III.9 and III.10) are now related by:

$$K_1 = \frac{a_H a_{HCO_3}}{a_{H_2CO_3}} = \frac{\gamma_H[H^+]\gamma_b[HCO_3^-]}{\gamma_a[H_2CO_3]} \tag{III.19a}$$

and

$$K_1' = \frac{[H^+][HCO_3^-]}{[H_2CO_3]} = \frac{\gamma_a}{\gamma_H\gamma_b}K_1 \tag{III.19b}$$

and, with respect to the second dissociation constant:

$$K_2 = \frac{a_H a_{CO_3}}{a_{HCO_3}} = \frac{\gamma_H[H^+]\gamma_c[CO_3^{2-}]}{\gamma_b[HCO_3^-]} \tag{III.20a}$$

and

$$K_2' = \frac{[H^+][CO_3^{2-}]}{[HCO_3^-]} = \frac{\gamma_b}{\gamma_H \gamma_c} K_2 \qquad\qquad (III.20b)$$

The definition also takes into account that instead of $[H^+]$, in reality the pH is being measured based on a series of buffer solutions. Therefore, in these equations $[H^+]$ is replaced by 10^{-pH}.

When comparing fresh and sea water, the differences in the first and second dissociation constants for carbonic acid – K_1 and K_2 for fresh water, and K_1' and K_2' for sea water – and their consequences, thereof will appear spectacular.

For practical reasons, the values of the dissociation constants are generally given as:

$$pK = -^{10}\log K \quad \text{or} \quad K = 10^{-pK} \qquad\qquad (III.21)$$

The K_0, K_1 and K_2 values for fresh water (ideal solution) and sea water as a function of the water temperature and the water salinity are discussed in the next sections, and are shown in Volume I of the UNESCO/IAEA series, Figs 9.1–9.4.

III.3.1   Ideal solutions

As far as the equilibrium constants are concerned, fresh waters can be considered as ideal solutions (extrapolated to zero ionic strength). Values for a temperature range of 0–40 °C and a salinity range of 0–40‰ are shown in Volume I, Figs 9.1–9.4, and in Table III.1 (left-hand shaded column) (see also: References and selected papers):

$$pK_0 = -2622.38/T - 0.0178471T + 15.5873 \text{ (Harned and Davis, 1943)}$$
$$(III.22)$$

$$pK_1 = 3404.71/T + 0.032786T - 14.8435 \text{ (Harned and Davis, 1943)} \quad (III.23)$$

$$pK_2 = 2902.39/T + 0.02379T - 6.4980 \text{ (Harned and Scholes, 1941)} \quad (III.24)$$

$$\ln K_w = 148.9802 - 13847.26/T - 23.6521 \ln T \text{ (Dickson and Riley, 1979)}$$
$$(III.25)$$

where the *absolute temperature* $T = t \, (°C) + 273.15 \, K$.

III.3.2   Seawater

The salt concentration of sea water is defined by the *salinity*, given in g/kg of sea water, or in ‰. Probably the best data have been reported by Millero and Roy (1997); these values for a temperature range of 0–40 °C and a salinity range of 0–40‰ are shown in Volume I, Figs 9.1–9.4 and in Table III.1 (right-hand bold column) (see also: References and selected papers).

$$\ln K_0' = -60.2409 + 9345.17/T + 23.3585 \ln(0.01T)$$
$$+ S[0.023517 - 0.023656(0.01T) + 0.0047036(0.01T)^2] \qquad (III.26)$$

Table III.1.   Apparent solubility and acidity (dissociation) constants of carbonic acid for various temperatures and salinities. (The values are according to Millero and Roy (1997) see: References and selected papers.)

$t(°C)$	S (‰)								
	0	5	10	15	20	25	30	35	40
$K_0 \cdot 10^2$									
0	7.691	7.499	7.295	7.112	6.934	6.761	6.592	**6.412**	6.252
5	**6.383**	6.223	6.067	5.916	5.754	5.610	5.470	**5.321**	5.188
10	**5.370**	5.236	5.105	4.977	4.842	4.721	4.603	**4.477**	4.365
15	**4.571**	4.457	4.345	4.236	4.121	4.018	3.917	**3.811**	3.715
20	**3.936**	3.837	3.741	3.648	3.556	3.459	3.373	**3.289**	3.199
25	**3.428**	3.342	3.258	3.177	3.090	3.013	2.938	**2.858**	2.786
30	**3.013**	2.938	2.864	2.793	2.723	2.655	2.582	**2.518**	2.449
35	**2.679**	2.612	2.547	2.483	2.415	2.355	2.296	**2.234**	2.178
40	**2.404**	2.344	2.286	2.223	2.168	2.113	2.061	**2.004**	1.954
$K_1 \cdot 10^7$									
0	**2.667**	4.667	5.433	5.984	6.412	6.761	7.047	**7.278**	7.464
5	**3.069**	5.420	6.353	7.015	7.534	7.962	8.299	**8.610**	8.851
10	**3.467**	6.194	7.295	8.072	8.690	9.204	9.638	**10.00**	10.30
15	**3.846**	6.966	8.241	9.162	9.908	10.52	11.04	**11.48**	11.86
20	**4.188**	7.727	9.183	10.26	11.12	11.83	12.45	**13.00**	13.46
25	**4.498**	8.433	10.09	11.32	12.33	13.18	13.90	**14.52**	15.07
30	**4.753**	9.099	10.96	12.36	13.52	14.49	15.35	**16.11**	16.75
35	**4.966**	9.705	11.80	13.37	14.69	15.81	16.79	**17.66**	18.41
40	**5.105**	10.26	12.56	14.32	15.81	17.06	18.20	**19.19**	20.09
$K_2 \cdot 10^{10}$									
0	**0.240**	1.291	1.879	2.355	2.773	3.155	3.508	**3.837**	4.150
5	**0.284**	1.600	2.339	2.951	3.491	3.981	4.436	**4.864**	5.272
10	**0.331**	1.950	2.877	3.639	4.325	4.943	5.534	**6.095**	6.607
15	**0.380**	2.350	3.491	4.436	5.284	6.081	6.823	**7.516**	8.185
20	**0.430**	2.793	4.188	5.346	6.397	7.379	8.299	**9.183**	10.05
25	**0.479**	3.289	4.966	6.383	7.656	8.872	10.02	**11.12**	12.19
30	**0.527**	3.828	5.834	7.534	9.099	10.57	11.99	**13.34**	14.66
35	**0.573**	4.426	6.792	8.831	10.72	12.50	14.22	**15.89**	17.50
40	**0.617**	5.070	7.870	10.28	12.53	14.69	16.79	**18.79**	20.75

$$pK_1' = 3670.7/T - 62.008 + 9.7944 \ln T - 0.0118\,S + 0.000116\,S^2 \qquad \text{(III.27)}$$

$$pK_2' = 1394.7/T + 4.777 - 0.0184\,S + 0.000118\,S^2 \qquad \text{(III.28)}$$

(K_0': Weiss (1974), K_1', K_2': Mehrbach *et al.* (1973), reported by Dickson and Millero (1987)).

The *salinity* values, S, are related to the originally used *chlorinity Cl*, that is, the concentration of chloride (plus bromide and iodide; also given in g/kg or ‰), by:

$$S = 1.80655\,Cl \qquad \text{(III.29)}$$

The solubility product of calcium carbonate differs for the two different crystalline types, calcite and aragonite. Volume I, Fig. 9.5 shows values at specific temperatures and salinities for calcite.

III.3.3 Brackish water

The large differences between K and K', that is, the large effect of salt concentrations on the acidity constants (Volume I, Fig. 9.4), result in an entirely different chemical character for fresh water and sea water. The acidity constants for fresh water with zero salinity and for sea water have been extensively studied experimentally. However, a problem remains for appropriate treatment of waters with varying low salt concentrations. In order to obtain correct K' values, the K values must be amended with the help of the Debye-Hückel theory, which is applicable at low concentrations. The treatment of non-ideal solutions is illustrated here by considering fresh waters with dissolved salts of less than about 400 mg/L.

A measure for the salt concentration is the *ionic strength*, I, of the water. This can be approximated by:

$$I \cong 2.5 \times 10^{-5} S \tag{III.30}$$

where S is the salt concentration in mg/L. The approximate values for the two acidity constants are then:

$$pK_1' = pK_1 - \frac{0.5\sqrt{I}}{1 + 1.4\sqrt{I}} \quad \text{and} \quad pK_2' = pK_2 - \frac{2\sqrt{I}}{1 + 1.4\sqrt{I}} \tag{III.31}$$

The acidity constant values published by Millero and Roy (1997) are derived from the given considerations on non-ideal solutions, taking into account a series of different compounds. Sets of data for various temperatures and salinities are presented in Table III.1.

Values for the solubility product of calcite ($CaCO_3$) (Eq. III.17) used in this volume are taken from Mucci (1983):

$$pK_{cal} = -^{10}\log K_{cal} = 171.9065 + 0.077993T - 2839.319/T - 71.595 \log T$$

$$+ (0.77712 - 0.0028426T - 178.34/T)S^{1/2}$$

$$+ 0.07711S - 0.0041249S^{3/2} \tag{III.32}$$

Values are shown in Volume I, Fig. 9.5 as functions of the water temperature for fresh- and sea water and as a function of salinity at 20 °C.

III.4 CARBONIC ACID CONCENTRATIONS

As previously mentioned, the differences between the acidity constants for fresh and sea water are considerable. This greatly effects the distribution of the carbonic acid fractions in natural waters. Examples of these are shown in the following sections.

The distribution of the dissolved carbonic acid species in pure water can be specified as a fraction of the total DIC. From Eqs III.6, III.9 and III.10, respectively, and Eqs III.18, III.19 and III.20 we obtain:

$$[H_2CO_3] = K_0 P_{CO_2} \tag{III.33}$$

$$[H_2CO_3] = \frac{[H^+]}{K_1}[HCO_3^-] \tag{III.34}$$

$$[CO_3^{2-}] = \frac{K_2}{[H^+]}[HCO_3^-] \tag{III.35}$$

so that

$$C_T = \left(\frac{[H^+]}{K_1} + 1 + \frac{K_2}{[H^+]} \right) [HCO_3^-] \qquad \text{(III.36)}$$

The fractional concentrations can now be given in terms of the total carbon content:

$$[HCO_3^-] = \frac{[H^+]K_1}{[H^+]^2 + [H^+]K_1 + K_1K_2} \times C_T \qquad \text{(III.37)}$$

$$[H_2CO_3] = [CO_2aq] = \frac{[H^+]^2}{[H^+]^2 + [H^+]K_1 + K_1K_2} \times C_T \qquad \text{(III.38)}$$

$$[CO_3^{2-}] = \frac{K_1K_2}{[H^+]^2 + [H^+]K_1 + K_1K_2} \times C_T \qquad \text{(III.39)}$$

and similarly for the apparent acidity constants K' in non-ideal solutions such as sea water.

The relative contributions of $[H_2CO_3]$, $[HCO_3^-]$ and $[CO_3^{2-}]$ to the total carbon content is shown in Fig. III.1 as a function of pH, for two different temperatures, for fresh water (at ionic strength $= 0$) and average sea water ($S = 35.0‰$, or $Cl = 19.37‰$).

As was anticipated, the acidity constants of carbonic acid change so rapidly with temperature and with the ionic strength of the solution that it results in a strong dependence of the distribution on both temperature and salinity.

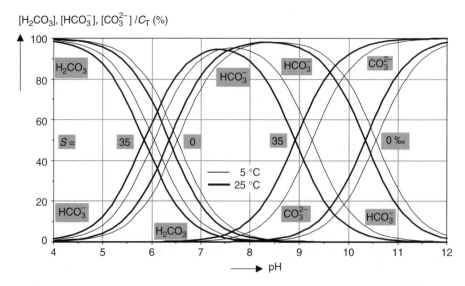

Figure III.1.   Distribution of carbonic acid fractions as percentages of the total carbon content, C_T. The values are calculated using Eqs III.37, III.38 and III.39 for temperatures of 5 and 25 °C, and for salinities of 0 and 35‰ as a function of the pH. Sea water has pH values around 8.2. However, the carbon distributions are shown for an (unrealistic) wider range of pH values to illustrate the dependence of the carbon distribution on salinity.

III.5 EXAMPLES FOR OPEN AND CLOSED SYSTEMS

This section gives a few examples of how to deal with the given equations. We refer to *closed systems* in which the dissolved carbonic acid fractions do not exchange with either the gaseous system (atmospheric or soil CO_2) or the solid phase ($CaCO_3$). In an open system the solution exchanges with the gaseous phase (= open to the gas phase) or the solid phase (open to $CaCO_3$). We deal with (i) a comparison with fresh and sea water, both in equilibrium with atmospheric CO_2, (ii) a groundwater sample for which $[H_2CO_3]$ corresponds to a P_{CO_2} value far exceeding the atmospheric concentration and which is brought into contact with the atmosphere at a constant temperature, (iii) water exposed to the atmosphere in the presence of calcium carbonate rock and (iv) the mixing of fresh water and sea water (in an estuary) under closed conditions.

III.5.1 Comparison of freshwater and seawater exposed to the atmosphere

Here we are dealing with a system which is open to the atmosphere (exchange with atmospheric CO_2) at a constant temperature. The latter is considered to be of infinite dimensions, so that P_{CO_2} is constant. The system is closed to $CaCO_3$, so that the carbonate alkalinity (= the concentration of positive metal ions) is constant. The specified conditions are shown in Fig. III.2.

Here we use the symbols:

$$a = [H_2CO_3] + [CO_2aq] \quad b = [HCO_3^-] \quad c = [CO_3^{2-}]$$

The values for the first and second (apparent) acidity constants are calculated from Eqs III.22 to III.28. Furthermore, Eqs III.37 to III.39 are applied for calculating the various dissolved fractions.

Starting with the known value for the atmospheric partial pressure of CO_2 the dissolved CO_2 concentration is obtained, and from that the pH-dependent other concentrations.

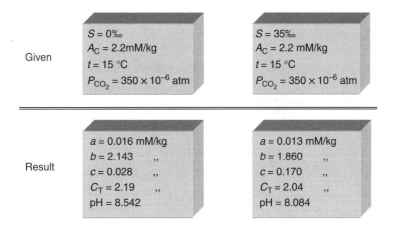

Figure III.2. Schematic representation for the conditions of fresh water (0‰) and sea water (35‰), both in equilibrium at a constant temperature with an atmosphere with constant CO_2 concentration (partial pressure). The upper part shows the given conditions and the lower part the resulting data.

III.5.1.1   *For freshwater*

$$a = [H_2CO_3] = K_0 P_{CO_2} = 0.04571(M/L \cdot atm)350 \times 10^{-6}(atm)$$
$$= 1.600 \times 10^{-5} \, M/L$$
$$b = [HCO_3^-] = (K_1/[H^+])a = (3.846 \times 10^{-7}/[H^+])a = 6.154 \times 10^{-12}/[H^+]$$
$$c = [CO_3^{2-}] = (K_1 K_2/[H^+]^2)a = (3.846 \times 10^{-7} \times 3.800 \times 10^{-11}/[H^+]^2)a$$
$$= 2.338 \times 10^{-22}/[H^+]^2$$

The carbonate alkalinity is known:

$$A_C = b + 2c = 6.154 \times 10^{-12}/[H^+] + 4.677 \times 10^{-22}/[H^+]^2 = 2.2 \times 10^{-3} \, M/L$$

From this quadratic equation $[H^+]$ and the pH can be obtained, resulting in:

$$[H^+] = 2.871 \times 10^{-9} \, M/L \quad \text{and} \quad pH = 8.542$$

The total carbon content of the water then is derived by inserting $[H^+]$ in the given equations:

$$C_T = a + b + c = 0.016 + 2.14 + 0.028 = 2.19 \, mM/L$$

III.5.1.2   *For seawater*

$$a = 0.03811(M/kg \cdot atm)350 \times 10^{-6}(atm) = 1.334 \times 10^{-5} \, M/kg$$
$$b = (11.480 \times 10^{-7}/[H^+])1.334 \times 10^{-5} = 15.314 \times 10^{-12}/[H^+]$$
$$c = (11.480 \times 10^{-7} \times 7.516 \times 10^{-10}/[H^+]^2) \times 1.334 \times 10^{-5}$$
$$= 1.151 \times 10^{-20}/[H^+]^2$$

Again the carbonate alkalinity is known:

$$b + 2c = 15.31 \times 10^{-12}/[H^+] + 2.302 \times 10^{-20}/[H^+]^2 = 2.2 \times 10^{-3} \, M/kg$$

Solving this quadratic equation results in:

$$[H^+] = 8.232 \times 10^{-9} \, M/kg \quad \text{and} \quad pH = 8.084$$

The total carbon content is:

$$C_T = 0.013 + 1.86 + 0.170 = 2.04 \, mM/kg$$

Comparing these results, as shown in Fig. III.1, reveals that in fresh water 98% of the carbonic species consists for of dissolved bicarbonate while the sea water, with a lower pH, still contains about 10% of dissolved carbonate ions.

III.5.2 System open for CO_2 escape and $CaCO_3$ formation

The second example (Fig. III.3) describes the chemical changes in a given fresh (ground)water body from which excess CO_2 escapes to the atmosphere, and in which $CaCO_3$ precipitates at saturation. In part (1) we indicate how the various carbon concentrations are calculated. In part (2) the solution loses CO_2 to the air (open to the atmosphere) at constant temperature until chemical equilibrium is reached between P_{CO_2} and the dissolved CO_2 fraction. Part (3) considers the presence of Ca^{2+} ions and a possible precipitation of calcite (open to the solid phase).

III.5.2.1 *Starting conditions*
We assume that the temperature, pH (resulting in $[H^+] = 10^{-pH}$) and the titration alkalinity (A) are known by measurement. The H^+ concentration will be denoted by the symbol h instead of $[H^+]$.

We can use the equations:

$$A = b + 2c + 10^{-14.34}/h - h, \text{ where the last two terms can be neglected}$$

$$b/a = K_1/h$$

$$c/b = K_2/h$$

Multiplication of the latter two relations gives:

$$c/a = K_1K_2/h^2 \quad \text{or} \quad c = (K_1K_2/h^2)a$$

With

$$A = b + 2c \text{ (for freshwater)}$$

this results in:

$$A = (K_1/h)a + (2K_1K_2/h^2)a$$

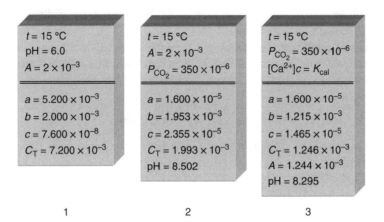

Figure III.3. Schematic representation of the numerical results from exposing a water sample with given conditions (upper part of 1) to the atmosphere (upper part of 2) and allowing precipitation of calcite (upper part of 3). The lower part of the boxes show the calculated data. All concentrations are in M/kg.

so that:

$$a = \frac{h^2}{hK_1 + 2K_1K_2}A \tag{III.40}$$

$$b = \frac{hK_1}{hK_1 + 2K_1K_2}A = \frac{h}{h + 2K_2}A \tag{III.41}$$

$$c = \frac{K_1K_2}{hK_1 + 2K_1K_2}A = \frac{K_2}{h + 2K_2}A \tag{III.42}$$

that is, three equations with three unknowns. At a temperature of 15 °C, a starting pH value of 6.0 and an alkalinity of 2 mM/L, the resulting concentrations are given in Fig. III.3, phase 1.

III.5.2.2   *Escape of CO_2*

If the water loses CO_2 in contact with the atmosphere, and by initially neglecting the presence of Ca^{2+} until equilibrium with the atmospheric P_{CO_2} (denoted by p) has been reached, we have:

$$a = K_0 p$$

$$b^2/(ac) = K_1/K_2$$

$$A \approx b + 2c$$

again three equations with three unknowns (bold type). With the additional starting values for the atmospheric CO_2 concentration (p), the concentration can now be calculated (Fig. III.3, phase 2).

III.5.2.3   *Precipitation of $CaCO_3$*

Apart from the known atmospheric P_{CO_2}, we have here the additional complication that calcite precipitates if the product of $[Ca^{2+}]$ and $[CO_3^{2-}]$ exceeds the solubility product given by Eq. III.32, so that Ca^{2+} decreases.

The amount of calcite (or Ca^{2+}) removed from the water, $\Delta[Ca^{2+}]$, equals the amount of carbon removed from the water, ΔC_T. We then have the following equations:

$$a_2 = K_0 p = a_1 = a$$

$$b_2/c_2 = b_1/c_1 = (K_1/K_2)a \quad \text{or} \quad c_2 = (c_1/b_1)b_2$$

$$A_2 - A_1 = [Ca^{2+}]_2 - [Ca^{2+}]_1 = C_{T2} - C_{T1} \quad \text{or} \quad \Delta[Ca^{2+}] = \Delta C_T$$

$$[Ca^{2+}]_2 c_2 = 10^{-pK_{cal}} \quad \text{or} \quad ([Ca^{2+}]_1 + \Delta C_T)c_2 = K_{cal}$$

$$\Delta C_T = (a + b_2 + c_2) - (a + b_1 + c_1) \quad \text{or} \quad \Delta C_T = b_2 + c_2 - (b_1 + c_1)$$

For reasons of simplicity we assume that the alkalinity is balanced by the Ca^{2+} ions alone, so that $[Ca^{2+}]_1 = 0.5A$ where subscript (1) refers to the values obtained by step 2, and (2) to the final values after precipitation of $CaCO_3$. These four independent equations, with four unknowns (bold type), can now be solved. The numerical result is shown in Fig. III.3.

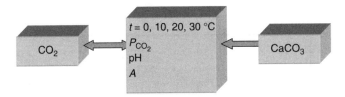

Figure III.4.   Schematic representation of an open system, consisting of a water mass in open contact with a CO_2 atmosphere and with solid $CaCO_3$.

Comparison of boxes 1 and 2 shows that:

1. The amount of CO_2 which has escaped to the atmosphere (= 143 mL STP) is in the order of the amount of dissolved CO_2 present at the start of the experiment.
2. The amount of calcite formed is 25.4 mg/L.

III.5.3   System exposed to CO_2 in the presence of $CaCO_3$

In the presence of carbonate rock fresh water may be exposed to a certain CO_2 pressure. The question now is how much $CaCO_3$ can be dissolved in equilibrium with the CO_2 atmosphere (Fig. III.4).

Instead of choosing the CO_2 pressure, the calculation procedure is easiest if we depart from the final pH the water will obtain. The alkalinity is defined by the electro-neutrality requirement:

$$A = 2Ca^2 + [H^+] = [HCO_3^-] + 2[CO_3^{2-}] + [OH^-] = \frac{2K_{cal}}{c} + h = b + 2c + \frac{K_w}{h}$$
$$(III.43)$$

Inserting Eqs III.33, III.34 and III.35 results in the equation:

$$\left(\frac{K_0K_1}{h} + 2\frac{K_0K_1K_2}{h^2}\right)P_{CO_2}^2 + \left(\frac{K_w}{h} - h\right)P_{CO_2} + 2\frac{K_{cal}h^2}{K_0K_1K_2} = 0 \qquad (III.44)$$

The solution of this equation and the resulting A values for four different temperatures are shown in Fig. III.5. The corresponding pH values are also indicated.

III.5.4   Closed system, mixing of freshwater and seawater

As an example of a closed system we discuss the mixing of fresh river water and sea water, such as occurs in an estuary (Fig. III.6). Calculating $^{13}\delta$ for the bicarbonate fraction of the brackish mixture from the $^{13}\delta$ values of bicarbonate in the end members (fresh and sea water) is not straightforward, because the mixing process of the mole fractions of bicarbonate is not conservative, that is, the dissociation equilibria of carbonic acid shift with the changes in pH, and the values of the acidity constants adjust with changes in the salinity of the water salinity.

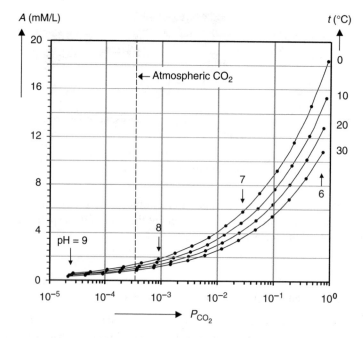

Figure III.5.   Schematic representation of an open system, consisting of a water mass in exchange with an infinite CO_2 reservoir and with carbonate rock. The alkalinities are calculated for four different temperatures. The corresponding pH values are indicated in the graph, as well as the atmospheric CO_2 partial pressure (P_{CO_2}).

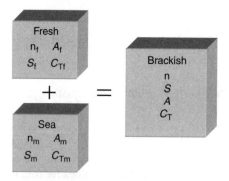

Figure III.6.   Schematic representation of the formation of brackish water as a mixture of fresh (river) and sea water. In this supposedly closed system the parameters shown in the figure are conservative.

The combined system for the two components is closed, however. As such, if the end members are denoted by f(resh) and m(arine), the final values for the conservative parameters, that is, the salinity, the alkalinity and the total carbon content are:

$$nS = n_f S_f + n_m S_m \quad \text{or} \quad S = (n_f/n)S_f + (n_m/n)S_m \tag{III.45}$$

where $n_f + n_m = n$, $S_f \approx 0\%o$ and $S_m \approx 35\%o$ (Fig. III.6).

Measurement of the salinity of the brackish water results in the (degree of) *brackishness* (X) of the water, here defined as the fraction of sea water (for sea water $X = 1$, for fresh water $X \approx 0$):

$$S = \frac{n_f}{n}S_f + \frac{n_m}{n}S_m = \left(1 - \frac{n_m}{n}\right)S_f + \frac{n_m}{n}S_m$$

or

$$X = \frac{n_m}{n} = \frac{S - S_f}{S_m - S_f} \approx \frac{S}{35} \tag{III.46}$$

Furthermore:

$$nA = n_f A_f + n_m A_m \quad \text{or} \quad A = (1 - X)A_f + XA_m \tag{III.47}$$

and

$$C_T = n_f C_{Tf} + n_m C_{Tm} \quad \text{or} \quad C_T = (1 - X)C_f + XC_m \tag{III.48}$$

In order to calculate $^{13}\delta_a$, $^{13}\delta_b$ and $^{13}\delta_c$, the carbonic acid fractions need to be obtained by solving the following four equations with four unknowns (in bold type):

$$A = b + 2c \qquad C_T = a + b + c$$
$$K_1 = hb/a \qquad K_2 = hc/b$$

where again $h = [H^+]$, $a = [CO_2]$, $b = [HCO_3^-]$ and $c = [CO_3^{2-}]$. K_1 and K_2 refer to the brackish water values at the brackish-water salinity.

A striking feature of estuaries is the non-linear behaviour of pH as a function of salinity: at a certain salinity the pH value shows a minimum. The position of this pH minimum strongly depends on the carbonate alkalinity ratio of the fresh and marine components (CA_f/CA_m) (see References and selected papers).

The consequence of this non-linear behaviour of pH, and thus of the carbonic acid fractions, is that $^{13}\delta$ for these fractions also does not obey conservative mixing (Section 5.3).

The only possible procedure for determining $^{13}\delta$ values for the dissolved carbonic components in water is by quantitatively converting the DIC into dissolved CO_2 by the addition of acid (such as H_3PO_4) to the water sample, and extracting the CO_2 from the water. Consequently, only $^{13}\delta_{DIC}$ can be obtained. If $^{13}\delta$ for the constituting fractions (a, b and c) are to be known, these should be calculated from the ^{13}C mass balance (cf. Section 3.1.1):

$$C_T\,^{13}R_{DIC} = a\,^{13}R_a + b\,^{13}R_b + c\,^{13}R_c = a\,^{13}\alpha_{a/b}\,^{13}R_b + b\,^{13}R_b + c\,^{13}\alpha_{c/b}\,^{13}R_b \tag{III.49}$$

or with $R/R_{std} = 1 + \delta$:

$$\begin{aligned} C_T\,^{13}\delta_{DIC} &= a\,^{13}\delta_a + b\,^{13}\delta_b + c\,^{13}\delta_c \\ &= a(^{13}\delta_b - ^{13}\varepsilon_{a/b}) + b\,^{13}\delta_b + c(^{13}\delta_b - ^{13}\varepsilon_{c/b}) \\ &= C_T\,^{13}\delta_b - a\,^{13}\varepsilon_{a/b} - c\,^{13}\varepsilon_{c/b} \end{aligned} \tag{III.50}$$

so that

$$^{13}\delta_{\text{DIC}} = {}^{13}\delta_{\text{b}} - (a/C_{\text{T}})^{13}\varepsilon_{\text{a/b}} - (c/C_{\text{T}})^{13}\varepsilon_{\text{c/b}} \qquad \text{(III.51)}$$

and similarly:

$$^{13}\delta_{\text{DIC}} = {}^{13}\delta_{\text{a}} - (b/C_{\text{T}})^{13}\varepsilon_{\text{b/a}} - (c/C_{\text{T}})^{13}\varepsilon_{\text{c/a}} \qquad \text{(III.52)}$$

$$^{13}\delta_{\text{DIC}} = {}^{13}\delta_{\text{c}} - (a/C_{\text{T}})^{13}\varepsilon_{\text{a/c}} - (b/C_{\text{T}})^{13}\varepsilon_{\text{b/c}} \qquad \text{(III.53)}$$

where the $^{13}\varepsilon$ values are given in Table 5.2.

References and selected papers

IN TABLE 5.2

[1] Mook, W.G., Bommerson, J.C. and Staverman, W.H., 1974. Carbon isotope fractionation between dissolved bicarbonate and gaseous carbon dioxide. *Earth and Planetary Science Letters* 22: 169–176.
[2] Vogel, J.C., Grootes, P.M. and Mook, W.G., 1970. Isotope fractionation between gaseous and dissolved carbon dioxide. *Zeitschrift Fur Physik* 230: 225–238.
[4] Thode, H.G., Shima, M., Rees, C.F. and Krishnamurty, K.V., 1965. Carbon-13 isotope effects in systems containing carbon dioxide, bicarbonate, carbonate and metal ions. *Canadian Journal of Chemistry* 43: 582–595.
[5] Emrich, K., Ehhalt, D. and Vogel, J.C., 1970. Carbon isotope fractionation during the precipitation of calcium carbonate. *Earth and Planetary Science Letters* 8: 363–371.

IN TABLE 5.4

[1] Majoube, M., 1971. Fractionnement en oxygen-18 et en deuterium entre l'eau et sa vapeur. *Journal de Chimie Physique* 68: 1423–1436.
[2] Friedman, I. and O'Neil, J.R., 1977. Compilation of stable isotope fractionation factors of geochemical interest. USGS Prof. Paper 440-KK.
[3] Brenninkmeijer, C.A.M., Kraft, P. and Mook, W.G., 1983. Oxygen isotope fractionation between CO_2 and H_2O. *Isotope Geoscience* 1: 181–190.
[5] Epstein, S., 1976. A revised oxygen paleotemperature scale (pers. comm.).

IN APPENDIX III

Dickson, A.G. and Riley, J.P., 1979. The estimation of acid dissociation constants in seawater from potentiometric titrations with strong base. I. The ion product of water-K_w. *Marine Chemistry* 7: 89–88.
Harned, H.S. and Davis Jr, R., 1943. The ionisation constant of carbonic acid in water and the solubility of carbon dioxide in water and aqueous salt solutions from 0° to 50°C. *Journal of American Chemical Society* 65: 2030–2037.
Harned, H.S. and Scholes, S.R., 1941. The ionisation constant of HCO_3^- from 0° to 50°C. *Journal of American Chemical Society* 63: 1706–1709.

Mehrbach, C., Culberson, C.H., Hawley, J.E. and Pytkowicz, R.M., 1973. Measurement of the apparent dissociation constants of carbonic acid in seawater at atmospheric pressure. *Limnology and Oceanography* 18: 897–907.

Millero, F.J. and Roy, R.N., 1997. A chemical equilibrium model for the carbonate system in natural waters. *Croatia Chemica Acta* 70: 1–38.

Mucci, A., 1983. The solubility of calcite and aragonite in seawater at various salinities, temperatures and one atmosphere total pressure. *American Journal of Science* 283: 780–799.

Weiss, R.F., 1974. Carbon dioxide in water and seawater: the solubility of a non-ideal gas. *Marine Chemistry* 2: 203–205.

The following list of 'references' cites papers that are of basic scientific and/or historic value. For a more complete list of articles on *isotope hydrology* the reader is advised to consult the series of UNESCO/IAEA Volumes.

Baertschi, P., 1976. Absolute ^{18}O content of Standard Mean Ocean Water. *Earth and Planetary Science Letters* 31: 341–344.

Bigeleisen, J., 1952. The effects of isotopic substitutions on the rates of chemical reactions. *Journal of Physical Chemistry* 56: 823–828.

Bigeleisen, J. and Wolfsberg, M., 1958. Theoretical and experimental aspects of isotope effects in chemical kinetics. *Advances in Chemical Physics* 1: 15–76.

Brenninkmeijer, C.A.M. and Mook, W.G., 1979. The effect of electronegative impurities on CO_2 proportional counting: an on-line purity test counter. In *Proceedings of the Ninth International Radiocarbon Conference*, Los Angeles, CA and La Jolla, 1976, University of California Press: 185–196.

Brenninkmeijer, C.A.M., Kraft, P. and Mook, W.G., 1983. Oxygen isotope fractionation between CO_2 and H_2O. *Isotope Geoscience* 1: 181–190.

Broecker, W.S. and Walton, A., 1959. The geochemistry of ^{14}C in fresh-water systems. *Geochimica et Cosmochimica Acta* 16: 15.

Craig, H., 1957. Isotopic standards for carbon and oxygen and correction factors for mass-spectrometric analysis of carbon dioxide. *Geochimica et Cosmochimica Acta* 12: 133–149.

Craig, H., 1961a. Standards for reporting concentrations of deuterium and oxygen-18 in natural waters. *Science* 133: 1833–1834.

Craig, H., 1961b. Isotopic variations in meteoric waters. *Science* 133: 1702–1703.

Dansgaard, W., 1964. Stable isotopes in precipitation. *Tellus* 16: 436–468.

Deines, P., 1980. The isotopic composition of reduced organic carbon. In P. Fritz and J.C. Fontes (eds), *Handbook of Environmental Isotope Geochemistry*, Vol. I: The Terrestrial Environment, Elsevier, Amsterdam, New York: 329–407.

De Wit, J.C., Van der Straaten, C.M. and Mook, W.G., 1980. Determination of the absolute D/H ratio of VSMOW and SLAP. *Geostandards Newsletter* 4: 33–36.

Emiliani, C., 1971. The amplitude of Pleistocene climatic cycles at low latitudes and the isotopic composition of glacial ice. In K.K. Turekian (ed.), *The Late Cenozoic Glacial Ages*, Yale University Press, New Haven, CT, London: 183–197.

Epstein, S. and Mayeda, T., 1953. Variations of O^{18} content of waters from natural sources. *Geochimica et Cosmochimica Acta* 4: 213–224.

Fontes, J.Ch., 1981. Paleowaters. In J.R. Gat and R. Gonfiantini (eds), *Stable Isotope Hydrology: Deuterium and Oxygen-18 in the Water Cycle*, IAEA Technical Report No. 210, Chapt. 12: 273–302.

Francey, R.J. and Tans, P.P., 1987. Latitudinal variation in oxygen-18 of atmospheric CO_2. *Nature* 327: 495–497.

Friedman, I., 1953. Deuterium content of natural water and other substances. *Geochimica et Cosmochimica* Acta 4: 89–103.

Fritz, P., 1981. River waters. In *Stable Isotope Hydrology: Deuterium and Oxygen-18 in the Water Cycle. Technical Report Series*. No. 210, IAEA, Vienna: 177–202.

Gat, J.R. and Tzur, Y., 1967. Modification of the isotope composition of rainwater by processes which occur before groundwater recharge. In *Proceedings of IAEA Symposium on Isotopes in Hydrology*, Vienna: 49–60.

Gat, J.R., 1996. Oxygen and hydrogen isotopes in the hydrological cycle. *Annual Review of Earth Planetary Sciences* 24: 225–262.

Gat, J.R., 1998. Modification of the isotopic composition of meteoric waters at the land-biosphere-atmosphere interface. In *Isotope Techniques in the Study of Environmental Change*, IAEA, Vienna: 153–163.

Geyh, M.A. and Wendt, I., 1965. Results of water sample dating by means of the model of Münnich and Vogel. In *Proceedings of International Conference on Radiocarbon and Tritium Dating*, Pullman, WA: 597–603.

Godwin, H., 1962. Half-life of radiocarbon. *Nature* 195: 984.

Gonfiantini, R., 1984. Stable isotope reference samples for geochemical and hydrological investigations. Report by Advisory Group's Meeting, Vienna, September 1983: 77 pp.

Hagemann, R., Nief, G. and Roth, E. 1970. Absolute isotopic scale for deuterium analysis of natural waters. Absolute D/H ratio for SMOW. *Tellus* 22: 712–715.

Ingerson, E. and Pearson, F.J.Jr, 1964. Estimating of age and rate of motion for groundwater by the C^{14}-method. In Recent Researches in the Fields of Hydrosphere, Atmosphere and Nuclear Chemistry. Maruzen Comp., Tokyo 263.

Inoue, H. and Sugimura, Y., 1985. Carbon isotope fractionation during the CO_2 exchange process between air and sea water under equilibrium and kinetic conditions. *Geochimica et Cosmochimica Acta* 44: 2453–2460.

Karlén, I., Olsson, I.U., Kållberg, P. and Killiçci, S., 1966. Absolute determination of the activity of two ^{14}C dating standards. *Arkiv för Geofysik* 6: 465–471.

Kaufman, S. and Libby, W.F., 1954. The natural distribution of tritium. *Physical Review* 93: 1337–1344.

Keeling, C.D., 1958. The concentration and isotopic abundances of atmospheric carbon dioxide in rural areas. *Geochimica et Cosmochimica Acta* 13: 322–334.

Keeling, C.D. and Whorf, T.P. Atmospheric carbon dioxide records for Mauna Loa and South Pole. http://cdiac.esd.ornl.gov/trends/co2

Keeling, C.D., Bacastow, R.B., Carter, A.F., Piper, S.C., Whorf, T.P., Heimann, M., Mook, W.G. and Roeloffzen, H., 1989. A three-dimensional model of atmospheric CO_2 transport based on observed winds: 1. Analysis of observational data. *Geophysical Monograph* 55: 165–236.

Kerstel, E.R.Th., van Trigt, R., Dam, N., Reuss, J. and Meijer, H.A.J., 1999. Simultaneous determination of the $^2H/^1H$, $^{17}O/^{16}O$, and $^{18}O/^{16}O$ isotope abundance ratios in water by means of laser spectrometry. *Analytical Chemistry* 71(23): 5297–5303.

Lal, D. and Peters, B., 1967. Cosmic-ray produced radioactivity on the Earth. In K. Sitte (ed.), *Handbuch der Physik 46*, Springer Verlag, Berlin: 551–612.

Levin, I. and Kromer, B., 2004. The tropospheric $^{14}CO_2$ level in mid-latitudes of the Northern Hemisphere (1959–2003). *Radiocarbon* 46(3): 1261–1272.

Libby, W.F., 1946. Atmospheric helium three and radiocarbon from cosmic radiation. *Physical Review* 69: 671–672.

Libby, W.F., 1965. Radiocarbon Dating (2nd ed.). University of Chicago Press, Phoenix Books, Chicago, IL, London: 175 pp.

Lucas, L.L. and Unterweger, M.P., 2000. Comprehensive review and critical evaluation of the half-life of tritium. *Journal of Research of the National Institute of Standards and Technology* 105: 541–549.

Mann, W.B., 1983. An international reference material for radiocarbon dating. *Radiocarbon* 25(2): 519–522.

Meijer, H.A.J. and Li, W.J., 1998. The use of electrolysis for accurate $\delta^{17}O$ and $\delta^{18}O$ isotope measurements in water. *Isotopes Environmental Health Studies* 34: 349–369.

Merlivat, L., 1978. Molecular diffusivities of $H_2^{16}O$, $HD^{16}O$, and $H_2^{18}O$ in gases. *Journal of Chemical Physics* 69: 2864–2871.

Merlivat, L. and Jouzel, J., 1979. Global climatic interpretation of the deuterium-oxygen 18 relationship for precipitation. *Journal of Geophysical Research* 84: 5029–5033.

Mook, W.G. and Koene, B.K.S., 1975. Chemistry of dissolved inorganic carbon in estuarine and coastal brackish waters. *Estuarine and Coastal Marine Science* 3: 325–336.

Mook, W.G., 1980. Carbon-14 in hydrological studies, In P. Fritz and J.Ch. Fontes (eds), *Handbook of Environmental Isotope Geochemistry 1A*, Elsevier Science Publishers, Amsterdam, New York: 49–74.

Mook, W.G., 1983. International comparison of proportional gas counters for ^{14}C activity measurements. *Radiocarbon* 25(2): 475–484.

Mook, W.G. and Streurman, H.J., 1983. Physical and Chemical Aspects of Radiocarbon Dating. PACT Publ.8 on ^{14}C and Archaeology: 31–55.

Münnich, K.O., 1957. Messung des ^{14}C-Gehaltes von hartem Grundwasser. *Naturwissenschaften* 44: 32.

Nier, A.O., 1950. A re-determination of the relative abundances of the isotopes of carbon, nitrogen, oxygen, argon, and potassium. *Physical Review* 77: 789–793.

Nydal, R. and Lovseth, K., 1983. Tracing bomb ^{14}C in the atmosphere 1962–1980. *Journal Geophysical Research* C 88: 3621–3642.

O'Neil, J.R. and Epstein, S., 1966. A method for oxygen isotope analysis of milligram quantities of water and some of its applications. *Journal of Geophysical Research* 71(20): 4955–4961.

Östlund, H.G. and Fine, R.A., 1979. Oceanic distribution and transport of tritium. In *Proceedings of IAEA Conference on the Behaviour of Tritium in the Environment*, San Fransisco: 303–312.

Pearson, F.J.Jr, 1965. Use of C^{13}/C^{12} ratios to correct radiocarbon ages of materials initially diluted by limestone. In *Proceedings of International Conference on Radiocarbon and Tritium Dating*, Pullman, WA: 357–366.

Rank, D., Adler, A., Aragúas Aragúas, L., Froehlich, K., Rozanski, K. and Stichler, W., 1998. Hydrological parameters and climatic signals derived from long-term tritium and stable isotope time series of the river Danube. In *Proceedings of IAEA Conference on Isotope Techniques in the Study of Environmental Change*, IAEA, Vienna: 191–205.

Roeloffzen, J.C., Mook, W.G. and Keeling, C.D., 1991. Trends and variations in stable carbon isotopes of atmospheric carbon dioxide. In *Proceedings of IAEA Conference on Stable Isotopes in Plant Nutrition, Soil Fertility and Environmental Studies*, IAEA, Vienna: 601–618.

Roether, W., 1967. Estimating the tritium input to ground water from wine samples: groundwater and direct run-off contribution to central European surface waters. In *Proceedings of IAEA Conference on Isotopes in Hydrology*, IAEA, Vienna: 73–90.

Schlosser, P., Shapiro, S.D., Stute, M., Aeschbach-Hertig, W., Plummer, N. and Busenberg, E., 1998. Tritium/^{3}He measurements in young groundwater. In *Proceedings of the Symposium Isotope Techniques in the Study of Environmental Change*, IAEA, Vienna: 165–189.

Shackleton, N.J., 1987. Oxygen isotopes, ice volume and sea level. *Quaternary Science Reviews* 6: 183–190.

Siegenthaler, U. and Oeschger, H., 1980. Correlation of ^{18}O in precipitation with temperature and altitude. *Nature* 285: 314.

Suess, H., 1955. Radiocarbon concentration in modern wood. *Science* 122: 415.

Urey, H.C., 1947. The thermodynamic properties of isotopic substances. *Journal of American Chemical Society*: 562–581.

Vogel, J.C., and Ehhalt, D., 1963. The use of carbon isotopes in ground-water studies, In *Proceedings of Conference on Isotopes in Hydrology*, IAEA, Vienna, 383–396.

White, J. From the website ftp.cmdl.noaa.gov/ccg/co2c13/flask/month, containing the CMDL CCGG concentration and isotopic data.

Zuber, A., 1983. On the environmental isotope method for determining the water balance components of some lakes. *Journal of Hydrology* 61: 409–427.

Literature

W.G. Mook (ed.)	*Environmental Isotopes in the Hydrological Cycle*	UNESCO/IAEA, Paris/Vienna (2000/2001)
W.G. Mook	Volume I: Introduction; theory, methods, review	
J.R. Gat, W.G. Mook and H.A.J. Meijer	Volume II: Atmospheric water	
K. Rozanski, K. Fröhlich and W.G. Mook	Volume III: Surface water	
M. Geyh	Volume IV: Groundwater	
K.-P. Seiler	Volume V: Man's impact on groundwater systems	
Y. Yurtsever (ed.)	Volume VI: Modelling Technical Documents in Hydrology, No. 39; IHP I–VI	
H. Moser	*Isotopenmethoden in der Hydrologie* (1980)	Gebr. Borntraeger, Berlin, Stuttgart
W. Rauert	(in German language) ISBN 3–443–01012–1	
P. Fritz J.Ch. Fontes	*Handbook of Environmental Isotope Geochemistry* ISBN 0–444–41781–8	Elsevier Science Publishing Company, Amsterdam, Oxford New York, Tokyo
	Volume 1. The Terrestrial Environment A (1980)	ISBN 0–444–41780–X
	Volume 2. The Terrestrial Environment B (1986)	ISBN 0–444–42225–0
	Volume 3. The Marine Environment A (1989)	ISBN 0–444–42764–3
F.J. Pearson e.a.	*Applied Isotope Hydrogeology*, a case study in Northern Switzerland (1991) ISBN 0–444–88983–3	Elsevier Science Publishing Company, Amsterdam, Oxford, New York, Tokyo
I. Clark P. Fritz	*Environmental Isotopes in Hydrogeology* (1997) ISBN 1–56670–249–6	Lewis Publishers, Boca Raton, New York

F. Gasse Ch. Causse	*Hydrology and Isotope Geochemistry* ISBN 2–7099–1377–1	Editions de l'Orstom, Paris
W. Kaess	*Tracing in Hydrogeology* (1998) ISBN 3–443–01013–X	Balkema
C. Kendall J.J. McDonnell	*Isotopes in Catchment Hydrology* (1998) ISBN 0–444–50155–X	Elsevier/North Holland Publishing Company, Amsterdam
E. Mazor	*Chemical and Isotopic Groundwater* *Hydrology – The applied approach* (1998) ISBN 0–8247–9803–1	Marcel Dekker Inc., New York
P.G. Cook A.L. Herczeg (ed.)	*Environmental Tracers in Subsurface* *Hydrology* (2000) ISBN 0–7923–7707–9	Kluwer Academic Publishers, Dordrecht
G. Friedlander J.W. Kennedy E.S. Macias J.M. Miller	*Nuclear and Radiochemistry* (1981) ISBN 0–471–86255–X	John Wiley & Sons, New York, Chichester, Brisbane, Toronto
G. Faure	*Principles of Isotope Geology* (1986)	John Wiley & Sons, New York, Chichester, Brisbane, Toronto

IAEA publications

IAEA CONFERENCE PROCEEDINGS

1963   Radioisotopes in Hydrology, Tokyo, 5–9 March 1963, IAEA, Vienna, 459 pp. (STI/PUB/71) (out of print)

1967   Isotopes in Hydrology, Vienna, 14–18 November 1966, IAEA, Vienna (in co-operation with IUGG), 740 pp. (STI/PUB/141) (out of print)

1970   Isotope Hydrology, Vienna, 6–13 March 1970, IAEA, Vienna (in co-operation with UNESCO), 918 pp. (STI/PUB/255) (out of print)

1974   Isotope Techniques in Groundwater Hydrology, Vienna, 11–15 March 1974, IAEA, Vienna, 2 volumes: 504 and 500 pp. (STI/PUB/373) (out of print)

1979   Isotope Hydrology (in 2 volumes), Neuherberg, Germany, 19–23 June 1978, IAEA, Vienna (in co-operation with UNESCO), 2 volumes of 984 pp. (STI/PUB/493) ISBN 92–0–040079–5 and ISBN 92–0–040179–1

1983   Isotope Hydrology, Vienna, 12–16 September 1983, IAEA, Vienna (in co-operation with UNESCO), 873 pp. (STI/PUB/650) ISBN 92–0–040084–1

1987   Isotope Techniques in Water Resources Development, Vienna, 30 March–3 April 1987, IAEA, Vienna (in co-operation with UNESCO), 815 pp. (STI/PUB/757) ISBN 92–0–040087–6

1992   Isotope Techniques in Water Resources Development, Vienna, 11–15 March 1991, IAEA, Vienna (in co-operation with UNESCO), 790 pp. (STI/PUB/875) ISBN 92–0–000192–0

1993   Isotope Techniques in the Study of Past and Current Environmental Changes in the Hydrosphere and the Atmosphere, Vienna, 19–23 April 1993, IAEA, Vienna, 624 pp. (STI/PUB/908) ISBN 92–0–103293–5

1995   Isotopes in Water Resources Management (in 2 volumes), IAEA, Vienna, 20–24 March 1995, IAEA, Vienna, 2 volumes: 530 and 463 pp. (STI/PUB/970) ISBN 92–0–105595–1 and 92–0–100796–5

1998   Isotope Techniques in the Study of Environmental Change, Vienna, 14–18 April 1997, IAEA, Vienna, 932 pp. (STI/PUB/1024) ISBN 92–0–100598–9

1999   Isotope Techniques in Water Resources Development and Management, 10–14 May 1999, IAEA, Vienna, CDRom (IAEA-csp-2/c) ISSN 1562–4153

SPECIAL IAEA SYMPOSIA

1967   Radioactive Dating and Methods in Low-Level Counting, Monaco, 2–10 March 1967, IAEA, Vienna, 744 pp. (STI/PUB/152) (out of print)

1979   Behaviour of Tritium in the Environment, San Fransisco, USA, 16–20 October
1978, 711 pp. (STI/PUB/498) ISBN 92–0–020079–6

1981   Methods of Low-Level Counting and Spectrometry, Berlin, Germany, 6–10 April
1981, IAEA, Vienna, 558 pp. (STI/PUB/592) (out of print)

IAEA REPORTS AND TECHNICAL DOCUMENTS (TECDOCS)

Environmental Isotope Data no.1–no.10: World Survey of Isotope Concentration in
Precipitation, Data from network of IAEA and WMO over period 1953–1991,
published 1969–1994.

Interpretation of Environmental Isotope and Hydrochemical Data in Groundwater Hydrology, *Proceedings of Advisory Group's Meeting*, Vienna, 27–31 January 1975, IAEA,
Vienna, 1976, 230 pp. (STI/PUB/429) ISBN 92–0–141076–X

Stable Isotope Standards and Intercalibration on Hydrology and Geochemistry,
R. Gonfiantini (ed.), Report on Consultants' Meeting, Vienna, 8–10 September 1976,
IAEA, Vienna, 1977.

Isotopes in Lake Studies, *Proceedings of Advisory Group's Meeting*, Vienna, 29 August–2
September 1977, IAEA, Vienna, 1979, 290 pp., ISBN 92–0–141179–0 (out of print)

Arid Zone Hydrology: Investigations with Isotope Techniques, *Proceedings of Advisory
Group's Meeting*, Vienna, 6–9 November 1978, IAEA, Vienna, 1980, 265 pp.
(STI/PUB/547) ISBN 92–0–141180–4

Palaeoclimates and Palaeowaters: A Collection of Environmental Isotope Studies, *Proceedings of Advisory Group's Meeting*, Vienna, 25–28 November 1980, IAEA, Vienna, 1981,
207 pp. (STI/PUB/621) ISBN 92–0–141083–2

Stable Isotope Hydrology Deuterium and Oxygen-18 in the Water Cycle, J.R. Gat and
R. Gonfiantini (eds), Monograph by Working Group, IAEA, Vienna, 1981, 340 pp.
(STI/DOC/10/210)

Guidebook on Nuclear Techniques in Hydrology, by Working Group IAEA, Vienna, 1983,
439 pp. (STI/DOC/10/91/2)

Stable Isotope Reference Samples for Geochemical and Hydrological Investigations,
R. Gonfiantini (ed.), Report by Advisory Group's Meeting, Vienna, 19–21 September
1983, IAEA, Vienna, 1984.

Stable and Radioactive Isotopes in the Study of the Unsaturated Soil Zone, *Proceedings
of Meeting on IAEA/GSF Programme*, Vienna, 10–14 September 1984, IAEA, Vienna,
1985, 184 pp. (TECDOC-357)

Isotope Techniques in the Study of the Hydrology of Fractured and Fissured Rocks,
Proceedings of Advisory Group's Meeting, Vienna, 17–21 November 1986, IAEA,
Vienna, 1989, 306 pp. (STI/PUB/790)

Stable Isotope Reference Samples for Geochemical and Hydrological Investigations, G. Hut
(ed.), Report on Consultants' Meeting, Vienna, 16–18 September 1985, IAEA, Vienna,
1987.

Isotopes of Noble Gases as Tracers in Environmental Studies, Report by Consultants'
Meeting, Vienna, 29 May–2 June, 1989, IAEA, Vienna, 305 pp. (STU/PUB/859) (out of
print) ISBN 92–0–100592–X

Use of Artificial Tracers in Hydrology, *Proceedings of Advisory Group's Meeting*, Vienna,
19–22 March 1990, IAEA, Vienna, 1990, 230 pp. (TECDOC-601)

C-14 Reference Materials for Radiocarbon Laboratories, K. Rozanski (ed.), Report on Consultants' Meeting, Vienna, 18–20 February 1981, IAEA, Vienna, 1991.

Statistical Treatment of Data on Environmental Isotopes in Precipitation, IAEA, Vienna, 1992, 781 pp. (STI/DOC/10/331)

Isotope and Geochemical Techniques applied to Geothermal Investigations, *Proceedings of Research Co-ordination Meeting*, Vienna, 12–15 October 1993, IAEA, Vienna, 1995, 258 pp. (TECDOC-788)

Reference and Intercomparison Materials for Stable Isotopes of Light Elements, *Proceedings of Consultants' Meeting*, Vienna, 1–3 December 1993, IAEA, Vienna, 1995. (TECDOC-825)

Manual on Mathematical Models in Hydrogeology, IAEA, Vienna, 1996, 107 pp. (TECDOC-910)

Manual for Operation of an Isotope Hydrology Laboratory, IAEA, Vienna, 1999. www. iaea.org/programmes/rial/pci/isotopehydrology/publications.shtml

Symbols and units

a	$[CO_2aq]$ = concentration of dissolved CO_2
a	year
^ya	activity ratio (^{y} mass number, e.g. ^{14}a)
aq	dissolved
A	(atomic) mass number
A	absolute (radio)activity (e.g. ^{14}A)
AMS	accelerator mass spectrometer
$\alpha_{l/v}$	fractionation factor (l rel. to v)
α_k	kinetic fractionation factor
α	alpha particle
b	$[HCO_3^-]$ = concentration dissolved bicarbonate
B	magnetic field
B	background counting rate
Bq	Becquerel = 1 disintegration $\cdot$ s^{-1}
β	beta particle
c	$[CO_3^{2-}]$ = concentration dissolved carbonate ions
°C	degree centigrade
C_T	concentration dissolved inorganic carbon
C_3	Calvin photosynthesis
C_4	Hatch-Slack photosynthesis
CAM	Crassulacean Acid Metabolism
C_i	Curie = 3.7×10^{10} dps
Cl	chlorinity (in g of chloride per kg of water = ‰)
d	deuterium excess of MWL
d	day = 8.6400×10^5s
dpm	disintegrations per minute
dps	disintegrations per second
D	diffusion constant/coefficient
DIC	dissolved inorganic carbon
DOC	dissolved organic carbon
$^x\delta$	relative isotope ratio (e.g. $^{13}\delta$) (defined from ^xR)
$^y\delta$	relative activity ratio (e.g. $^{14}\delta$) (defined from ^yA or ^ya)
E	energy
E_B	binding energy
EC	electron conversion
ε	fractionation (constant) (enrichment/depletion)

ε_k	kinetic fractionation (constant)
f	fraction
F	force
g	gram
GM	Geiger-Müller counter
γ	gamma 'particle'/radiation
h	relative humidity
h	hour
I	electric current
IAEA	International Atomic Energy Agency
IRMS	isotope ratio mass spectrometer
J	Joule
keV	kiloelectronvolt $= 10^3$ eV
K	equilibrium/acidity constant
K	degree Kelvin
LSS	liquid scintillation spectrometer
λ	(radioactive) decay constant
m	mass
m	metre
min	minute
mol	symbol for mole
mole	number of grams equal to molar weight
M	molar weight, mole
MeV	millionelectronvolt $= 10^6$ eV
MS	mass spectrometer
MWL	meteoric water line
μ	reduced mass
n	neutron
N	neutron number
N	amount
NBS	National Bureau of Standards
NIST	National Institute of Standards and Technology, USA
ν	neutrino
ν	frequency
Ox	oxalic acid (^{14}C standard)
p	pressure
p	proton
pH	$= -^{10}\log[H^+]$
pCi	picoCurie $= 10^{-12}$Ci
pMC	per cent Modern Carbon
P	probability
PDB	PeeDee Belemnite
PGC	proportional gas counter
q	partition function
q	electric charge
Q	nuclear reaction energy
r	radius

^xR	isotope ratio (^{x} mass number, e.g. ^{13}R)
s	second
s	slope of MWL $= 8$
sp	spallation
S	salinity (in g of salt per kg of water $= \text{‰}$)
SLAP	standard light Antarctic precipitation
SMOW	standard mean ocean water
STP	standard temperature and pressure ($0\,°C$, 1 atm)
σ	standard deviation
Σ	total inorganic carbon concentration $= C_T$
t	time
t	temperature (in $°C$)
T	absolute temperature (in K)
$T_{1/2}$	half-life
TU	tritium unit $\equiv [^3H]/[^1H] = 10^{-18} = 0.118$ Bq/L
θ	exponent in fractionation factor ratio for $(\Delta M = 2)/(\Delta M = 1)$
τ	mean life
v	velocity
V	volume
V	volt
VPDB	Vienna-PDB
VSMOW	Vienna-SMOW
Z	atomic number

Constants

a	year $= 3.1558 \times 10^7$ s
amu	atomic mass unit $= 1.66054 \times 10^{-27}$ kg
c	velocity of light (in vacuum) $= 2.997925 \times 10^8$ m $\cdot$ s^{-1}
cal	calorie $= 4.184$ J
e	elementary/electron/proton charge $= 1.60218 \times 10^{-19}$C
eV	electronvolt $= 1.60218 \times 10^{-19}$ J
g	acceleration of free fall $= 9.80665$ m $\cdot$ s^{-2}
h	Planck constant $= 6.62608 \times 10^{-34}$ J $\cdot$ s
J	Joule $= 0.2390$ cal
k	Boltzmann constant $= 1.38054 \times 10^{-23}$ J/K
m_e	electron mass $= 9.10939 \times 10^{-31}$ kg
m_n	neutron mass $= 1.67493 \times 10^{-27}$ kg
m_p	proton mass $= 1.67262 \times 10^{-27}$ kg
M/E eq.	mass/energy equivalence: 1 amu $\equiv 931.5$ MeV
N_A	Avogadro constant $= 6.02214 \times 10^{23}$ mol^{-1}
π	$= 3.1415926535\ldots$
R	gas constant $= 8.31451$ J $\cdot$ K^{-1} $\cdot$ mol^{-1}
T	thermodynamic temperature $= t(°C) + 273.15$ K
V_m	molar volume ($= 22.41$ L $\cdot$ mole^{-1} at STP)

Index

IAH International Contributions to Hydrogeology

eBooks – at www.eBookstore.tandf.co.uk

A library at your fingertips!

eBooks are electronic versions of printed books. You can store them on your PC/laptop or browse them online.

They have advantages for anyone needing rapid access to a wide variety of published, copyright information.

eBooks can help your research by enabling you to bookmark chapters, annotate text and use instant searches to find specific words or phrases. Several eBook files would fit on even a small laptop or PDA.

NEW: Save money by eSubscribing: cheap, online access to any eBook for as long as you need it.

Annual subscription packages

We now offer special low-cost bulk subscriptions to packages of eBooks in certain subject areas. These are available to libraries or to individuals.

For more information please contact webmaster.ebooks@tandf.co.uk

We're continually developing the eBook concept, so keep up to date by visiting the website.

www.eBookstore.tandf.co.uk